A Field Guide to

AUTOMOTIVE
TECHNOLOGY

ED SOBEY

D1089285

CHICAGO
REVIEW
PRESS

Library of Congress Cataloging-in-Publication Data
Sobey, Edwin J. C., 1948–
    A field guide to automotive technology / Ed Sobey.
    p. cm.
    Includes index.
    ISBN 978-1-55652-812-5
    1. Automobiles—Popular works. 2. Mechanics—Popular works. I. Title.

TL146.5.S63 2008
629.2—dc22

                                        2008046620

Cover and interior design: Joan Sommers Design
Cover photos: Front cover, clockwise from top left: Speedometer, Wing Mirror photos
courtesy of author; Gray Metallic Sports Car photo by Stockxpert; Car Parts Suspension
photo by César Gutiérrez Wong/iStockphoto; Gearbox photo by Pawel Zawistowski/
Stock.xchng; Beneath the Hood photo by Jan Roger Johannesen/Stock.xchng. Back cover,
top to bottom: Flywheel, Power Steering, Brakes, Spoiler photos courtesy of author.
Photo on page 28: © Smokey Combs

Published by Chicago Review Press, Incorporated
814 North Franklin Street
Chicago, Illinois 60610
ISBN 978-1-55652-812-5
Printed in the United States of America
5 4 3 2 1

To all of those greasy knuckled people who tinker and think of better ways to do things.

CONTENTS

Acknowledgments 6

1 IGNITION! 7
A Brief History of Wheeled
  Vehicle Technology 7
How Cars Work 10

2 ON THE CAR 13
Antenna, AM/FM 14
Antenna, Citizens Band Radio (CB) ... 15
Antenna, OnStar 16
Antenna, Satellite Radio 17
Autopark and Back-Up
  Proximity Systems 19
Bumper 21
Convertible Top 22
Headlights 24
Heating Plug 26
Hubcaps and Spinners 28
License Plate 29
Spoiler 30
Windshield 32
Windshield Wipers 33
Wing Mirror 35

3 INSIDE THE CAR 37
Air Bag 38
Air Conditioning 40
Automatic Windshield Wipers 42
Auxiliary Heater 43
Brake Light 44
Brake Pedal 45
CD Player 47
Child Car Seat 48
Cruise Control 49
Defrost System Control 51
DVD Player 52
Flares (Fusee) 53
Four-Wheel-Drive Shifter 54
Fuel Gauge 56
Fuses 57
Glove Box 59
Global Positioning System (GPS) 60
Hand-Cranked Window 62
Heater 63
Key Fob 64

Odometer 66
Parking Brake 68
Power Window 69
Radar Detector 70
Radio 72
Rearview Mirror 74
Seat Belt 76
Speedometer 78
Steering Wheel 79
Tachometer 81
Temperature Gauge 82
Tire Pressure Gauge 83
Toll Transponder 84
Turn Indicator 85

4 UNDER THE CAR 87
Brakes 88
Catalytic Converter 89
Coil Spring 91
Constant Velocity Joint Boot 92
Differential 93
Gas Tank 95
Jack 96
Leaf Springs 97
Muffler 98
Rack and Pinion Steering 100
Resonator 101
Roll Bar (A.K.A. Anti-Roll Bar or
  Sway Bar) 102
Shock Absorber 103
Springs 104
Struts 105
Tailpipe 106
Tie Rod 107
Tires 109
Transfer Case 112
Universal Joint (U-Joint) 113
Wheel 114
Wheel Clamp (or Denver Boot) 115

5 UNDER THE HOOD 117
Internal Combustion Engines 117
Electric Motors 119
Hybrid Motors 120
Air Filter 122

Alternator 123
Battery 125
Brake Cylinder (or Master Cylinder) . . 127
Coil 129
Dipstick 130
Distributor 131
Fan 132
Horn 133
Oil Filter 135
Power Steering 137
Radiator 139
Spark Plug 141
Starter 142
Thermostat 144
Transmission 145
Turbocharger 147
Water Pump 149
Windshield Cleaning System 150
Windshield Wiper Motor 151

6 OFF-THE-ROAD
PASSENGER VEHICLES 153
Amphicar and Aquada 154
All-Terrain Vehicle (ATV) 156
DUKW 157
Golf Cart 158
Snowcat 159
Snowmobile 160

7 HUMAN-POWERED
VEHICLES 163
Bicycle Escalator 165
Bike Suspension System 167

Brakes 168
Derailleur 169
Quick-Release Hub 171
Pedicab or Cycle Rickshaw 172
Unicycle 173
Kick Sled 175
Scooter 176

8 MOTORCYCLES 177
Brakes 179
Carburetor 180
Engine 182
Exhaust System 184
Foot Controls 185
Gasoline Tank 186
Hand Controls 187
Oil Tank 188
Radiator 189
Shock Absorbers 190
Sidecar 191
Segway 192

9 BUSES 195
Bus Tracking System 196
Fare Box 197
Outside the Bus 199
Inside the Bus 201
Trolley 203

Index 205

ACKNOWLEDGMENTS

To help me write this book I recruited an automotive brain trust from among my friends. Laine Boekelman gave me a primer on motorcycles. What Laine didn't cover, Willie Sato did. Willie even washed his motorcycle before I arrived so it would look nice in the photographs.

Doug Chase, who has his own business of building race cars, answered lots of questions.

John Blake, a professional mechanic, allowed me into his garage to watch him repair cars and hear his explanations of how various parts work. In a life with no spare time, John gave me some. Thank you.

Ed Pfeiffer took me on a tour of a bus barn, inside a few buses, and around the trolleys. That was fun. Dan Overgaard with King County Metro Transit provided great information on bus tracking.

Thanks go to Rich Sidwa who again provided many photographs, as he has for earlier books. We stood outside on a cold and rainy day taking photos. Rich also is quite knowledgeable about cars and was able to steer me straight.

Bike escalator photos were provided by Jarle Wanvik. He is the creator of the escalator (www.trampe.no) and we hope he will be successful in getting more cities to adopt them. Russ Noe provided photos of sidecars. The photo of the Amphicar was taken by Ed Price, who is an avid amphibian-car enthusiast. Stan Wolfson of Clancy Systems in Denver provided the photo of the Denver boot. Smokey Combs provided the image of the wheel spinners. Thanks to all.

IGNITION!

Why gas-guzzling cars? Why is our transportation dominated by four wheels powered by a gasoline-snorting engine?

People have been using wheels for nearly 6,000 years. The invention of the wheel probably occurred many times in many places and no event of inception was recorded. At first wheels were powered by the people who made them. Hitching animals to move carts started around 4,000 years ago.

Animals work well pulling people and cargo, but have some serious drawbacks. By the 1880s, New York City had to dispose of 15,000 dead horses that had been left in the streets each year. The city was also engaged in the business of collecting and disposing of 20 tons of horse manure every day. Watching a car belch its exhaust may annoy us, but picture following a team of horses clopping down the street soon after they had eaten their oats. There were serious health concerns about the piles of rotting manure left scattered throughout the city and the accompanying flies. People also complained of the din of iron horseshoes hitting the paving; the noise was so loud that people had trouble talking to one another on the streets. Life for the horses wasn't so great either. Life expectancy of a working horse was about four years, and many were mistreated.

The steam engine changed everything. The concept for steam power had been around since the first century—Hero's Engine, called an aeolipile, was a working steam engine but an impractical one. In the 18th century tinkers started applying new technologies of metallurgy to containing and controlling the power of steam. James Watt made a huge contribution by building an improved steam engine with an external condenser. This innovation thrust steam power into the realm of practicable technology.

The first steam vehicle in the United States was a strange device made by inventor Oliver Evans. Evans's contraption, named the Orukter Amphibolos, could run on land or water. It was designed as a motorized river dredge that could travel over land to get to the dredge site. The dredge was probably never used but inspired generations of early American inventors to try steam power.

Steam power for vehicles was popular well into the 20th century. In 1906 driver Fred Marriott set a land speed record of 121 mph in the Rocket, a steam-powered race car. The Rocket set a new record of 132 mph the following year before crashing.

But steam wasn't alone as a power source for vehicles. Scientific discoveries had led to practical applications for electricity, including the electric motor. By the end of the 19th century, car companies were making both steam and electric vehicles. And a few companies were starting to use the newly invented internal combustion engines.

At the start of the 20th century, internal combustion automobiles ran a distant third behind those powered by steam or electric engines. Electric cars especially were safer to use, provided a smoother and quieter ride, and were easier to operate. Industry experts predicted the demise of the gasoline engine as it was noisy and unreliable, and it delivered an uncomfortable ride. The only certainty in the future of vehicle engines seemed to be that people would be driving cars powered by either steam or electricity.

Today, as electric engines are resurging amid the green revolution and fuel-cost consciousness, it's hard to imagine how electric cars lost market share to gasoline. But internal combustion proponents worked steadily to reduce their engines' drawbacks.

Gasoline engines operate in a relatively narrow range of rotational speeds. While this is not a problem for a lawn mower that chomps away at a steady rate, it is a big problem in powering a car from zero to 60 miles per hour. The invention of the transmission (and much later the automatic transmission) made gasoline and diesel engines competitive.

Starting a gasoline engine was a difficult and dangerous job until Charles Kettering's invention of the automatic starter removed that liability. Kettering also invented the electric ignition system, leaded gasoline (now outlawed due to concerns of lead in the environment), four-wheel brakes, and safety glass.

While gasoline-powered cars became easier to operate, steam remained complex. Although a well-run steam car could keep up with both electric and gasoline cars, steam became increasingly more impractical by comparison.

Initially, engine-powered vehicles were toys for the wealthy. Electric and steam-powered cars never broke out of that mold. Electrics were especially expensive to purchase, although they were cheaper to operate than gasoline—the same as today. The companies that made steam and electric cars focused on serving the limited customer base of the rich. Utility took a backseat to class appeal.

When Henry Ford's grand experiment with mass production took shape, the cost of gasoline cars plummeted. He succeeded in his goal to make cars affordable for the working class. Now people could use cars as practical transportation and not just for weekend picnics. By 1917 the race for dominance had been won by gasoline proponents. Although there were some 50,000 electric-powered cars in the United States that year, there were 70 times more gasoline-powered cars.

Ford succeeded because his engineers were successful in solving the problem of production. The 1908 Model T was so successful that Ford had trouble keeping up with demand in his traditional assembly plants. The Model T ran well on the unpaved roads of America and it ran with little need for expert maintenance—which is good, because little was available. Since Ford was selling every car they could manufacture, they focused on increasing production. It took Ford six years to develop the moving assembly line, which was launched in 1914.

The combination of technological innovations and the economic rise of the middle class ushered in the age of the internal combustion machine. Steam and electric vehicles were soon forgotten.

Trucks followed cars by a few years. The Winton Motor Carriage Company made the first in 1898. Unlike cars, trucks caught on slowly. There wasn't a ready market for them. Horse-drawn wagons were far less costly and were more efficient in some industries. In the home delivery of milk, for example, the horse would move down the street independent of the driver who was walking to leave bottles on the front porches of customers. No gasoline-powered truck could operate unattended like a horse-drawn wagon. And although gasoline-powered trucks could travel farther faster, most deliveries were local and horses worked well for those. Also, the largest businesses had the most money invested in the existing technology—horses and the tack they required—and were protective of that investment and resistant to new technology.

The need to haul more heavy goods farther coupled with the addition of the trailer lead to increased sales of trucks. But it was during World War I that trucks proved reliable. Following the war the road systems in the United States and Europe were improved, making trucks even more practical. And each new innovation in engine technology, suspension, and steering made trucks the practical choice.

Today we take gasoline-powered cars and trucks for granted. Some 45 million are built worldwide every year. But is the end in sight? Will other more environmentally friendly engines take its place?

HOW CARS WORK

Explosions! Thousands of explosions every minute of operation power internal combustion engines. Squirt one part of fuel and 15 parts of air into a closed cylinder, add an electric spark, and there will be an explosion.

Explosions are rapid chemical reactions that release tremendous amounts of energy, mostly as heat. The gases created in the explosion expand rapidly, increasing the pressure inside the cylinder and driving a moveable piston down the cylinder.

A crankshaft converts the up and down motion of several pistons into rotary motion that powers the wheels. But to get to the wheels, the kinetic energy must transfer through a transmission that trades engine speed for torque, or turning power, through a series of gears. Moving torque from the transmission to the wheels requires complex mechanical systems that have great variety in design.

Is this all? Not at all. There is much more to how a car works. But this is a start. Now go look at your car—ask yourself what each part does, and if you don't know the answer look it up in the following pages.

IT'S ELEMENTAL

What chemical elements is your car made of? By weight, metals predominate. Average cars carry about one ton of iron. But after that heavy load, the list of metals slims down. Aluminum comes in at about 250 pounds. Copper and silicon (mostly in glass) weigh in at nearly 50 pounds. Cars have about as much lead (in the battery) as zinc (for rust protection): about 20 pounds. Cars have less than 20 pounds of manganese, chromium, nickel, and magnesium.

ON THE CAR

MUCH OF YOUR CAR'S TECHNOLOGY is hidden beneath the metal and plastic body or hood. But some equipment cannot be hidden or protected inside the car. In some cases designers blend the machines into the car's body so you don't notice them. Others are themselves design elements and some pop out from hidden recesses when needed.

Antenna, AM/FM

BEHAVIOR

It wiggles in the wind as you drive at highway speeds, showing patterns of standing waves. It also receives the radio signals that bring you news, sports, music, and way too many commercials. As if that weren't enough, it also provides a perch for antenna balls.

HABITAT

On most cars it is the stiff wire that rises vertically from just in front of the windshield on the passenger's side or on the rear fender on the driver's side.

HOW IT WORKS

Antennas are tuned to receive electromagnetic radiation within certain frequency bands. Note their similarity to tiny antenna on old cell phones. (Newer cell phones, operating at even higher frequencies, have smaller antenna that fit inside the hand unit.) AM and FM radio stations broadcast at low frequencies and large antennas are needed to receive those signals at these frequencies.

To transmit an AM signal the ideal antenna is huge. Hence, AM radio stations have very tall towers and long antenna. FM stations, which operate at higher frequencies, need shorter transmit antennas. But both types of stations have transmit antennas many times larger than the antenna on your car. Driving around with a 100-foot-tall antenna just won't work, so the transmitted signals are strong enough that the less than optimum height antenna on your car still receives radio signals.

INTERESTING FACTS

Radio antennas had been mounted in the cloth roofs of cars until the advent of steel roofs for cars in 1934. The new roofs reflected and blocked radio waves, so engineers experimented with placing antenna elsewhere, eventually settling on the favored location behind the hood.

Antenna, Citizens Band Radio (CB)

BEHAVIOR

Long and lanky, the CB antenna bends and sways as the pickup truck it's attached to accelerates. It pulls radio from the electromagnetic atmosphere and sends back replies: "That's a ten-four, good buddy."

HABITAT

Long CB antennas are often mounted on a bumper to keep them low enough to fit into garages. Shorter CB antennas are mounted on the roof or on side mirrors of trucks.

HOW IT WORKS

In the United States, citizens band radio operates in the band of frequencies around 27 MHz. Within this band of frequencies 40 channels are designated for CB use. CB users can select any of the channels to use. One channel, 16, is reserved for meeting other users and agreeing which other (lower-traffic) channel to use for conversation.

The radio wave at 27 MHz is 11 meters long. To best capture that signal, the antenna needs to be either one half or one quarter of the wavelength. One half of 11 meters would be too long to use on cars and trucks, so the preferred antenna length is one quarter of 11 meters, or 2.7 meters. That is still quite tall, so the antenna is often mounted on the lowest spot possible—the bumper. To protect the car from being scratched by the antenna as it moves, the antenna is often outfitted with a tennis ball that can bounce against the car.

In many cases, the 2.7-meter antenna would still be too long, so a loading coil is inserted into a shortened antenna. The coil improves reception on shorter than quarter-length antenna. A loading coil can be located anywhere along the length of the antenna, but is often near its base.

Antenna, OnStar

BEHAVIOR
This GM system is a subscription service that can provide vehicle tracking (for stolen cars), emergency response (notifying authorities of an emergency and its location), and other communications. Newer versions of OnStar automatically contact emergency services if the vehicle is involved in a serious accident. Some systems allow police to shut off the car's engine if it has been reported stolen.

HABITAT
The antenna is usually found on the back of the roof, in the center. It often has a distinctive shark-fin shape, but other shapes are used as well.

HOW IT WORKS
OnStar uses cellular telephone systems to communicate. Emergencies are handled out of two call centers operated around the clock: one in Charlotte, North Carolina, and the other in Oshawa, Ontario.

The system has a diagnostic system to sense problems, such as impacts that suggest a collision. When an impact is recorded, the system communicates to the operation centers by cell phone service provided by the three major cell phone companies in the United States.

The service includes a built-in car phone. The driver can make and receive calls without picking up a phone. Calls are made hands-free.

Antenna, Satellite Radio

BEHAVIOR

The advantage of having satellite radio reception is being able to drive completely across the country and never having to change your radio dial. Or being able to listen to every NFL football game regardless of where you are. Satellite radio delivers dozens of music and entertainment channels, plus sports, news, and traffic information nearly everywhere in the United States, including southern Alaska. Television service for backseat viewers will soon be available by satellite radio.

HABITAT

These antennas can take one of several shapes. Most common is a vertical wire sheathed in plastic about a foot long that has a plastic base attached to the car. Another model added after market is a small plastic box with wires that can be fed into the trunk. All are mounted on the roof or other parts high enough to receive signals from overhead.

HOW IT WORKS

The two satellite companies operating in the United States, Sirius and XM Satellite Radio, merged in February 2007. Because the two companies use incompatible technology, they will have redundant equipment and services until they introduce radio receivers that can receive signals from both systems. The combined company has seven satellites in space plus one spare for each of the two technologies.

XM satellites are geostationary, while Sirius satellites are geosynchronous. A geostationary satellite revolves around the Earth at the same rate that the Earth is spinning, so it stays over the same point relative to Earth. These are located above the equator. Geosynchronous satellites return to the same location above Earth at the same time every day. Having multiple geosynchronous satellites allows the radio company to have one above the center of the United States at all times. This reduces the number of repeaters they need on the ground. The spares are kept on hand to replace a satellite should it fail.

In addition to the satellites, there is a network of ground repeaters that fill in the signal in locations that don't have good reception from the satellite. A typical U.S. city might have 20 repeaters. XM operates about 800 repeaters in the United States.

The satellites broadcast (and the repeaters repeat) a signal within the frequency band centered at 12.5 MHz. They broadcast on two carrier waves within the 12.5 MHz band and use four other bands to repeat the signal. A complex system allows one signal to fill in for another.

The visible receivers catch the radio signals from either satellite or ground repeater, filter out unwanted radio signals, and amplify the signal. The second component of the system decodes the radio signals and lowers the frequency of the signals so the car radio can play the songs.

The name Sirius comes from the name of the brightest star in the night sky.

Autopark and Back-Up Proximity Systems

BEHAVIOR

For the parking-impaired (like me), the autopark or self-park drives the car into tight parallel parking spots. They also assist with backing into a parking space. Less sophisticated systems provide distance warnings as the cars backs up.

HABITAT

Some of the electronics are housed in the dashboard, but the controlling computer is mounted inside the trunk. Sensors are mounted in the front and rear bumpers and on the fenders.

HOW IT WORKS

Several sensors detect other cars and estimate the distance to them. They also estimate how much space is available in the parking space and the distance to the curb. Data is fed into a computer that calculates the optimal steering angles and then controls the car's steering.

System sensors are energized when the driver puts the transmission in reverse. The computer alerts the driver when to shift gears and when to stop. The driver controls the car's speed, by pressing on the brake pedal, and the transmission—forward and reverse. The computer controls the steering.

Sensors use ultrasound sonar to measure the distance to any objects. Sonar systems measure the length of time between the sending of a pulse and receiving a reflection of the pulse. The longer the time, the farther away the object is.

While backing up, the sensors trigger a warning beep played on a piezoelectric speaker inside the car. As the car gets closer to another vehicle or other object behind it, the pace of the beeps increases.

Some systems also have a video screen that illustrates how close the car is getting to the object behind it. More elaborate systems, like those found on some models of Lexus, have a video camera to show what is behind the car. The video screens have touch screen controls so the driver can tell the system where he or she wants to park.

These systems are new and only a few car models have them. They seem to be popular with car buyers, so expect to see more models available soon.

Bumper

BEHAVIOR
They don't do much, except when you drive too far into a parking space. Then they alert you with a bump and a noise that tells you, "Oh, no."

HABITAT
They protrude beyond the car, both stem and stern, ostensibly to protect the more expensive components of the car from collisions.

HOW IT WORKS
The idea is sound: put a sacrificial steel bar that can withstand the bruises of everyday traffic to protect the more valuable fenders, grill, hood, and other expensive parts. Over time, however, bumpers have become refined and, in the process, less able to do their assigned task.

Fiberglass has replaced steel for bumpers and their role has changed from useful protection to ornamentation. However, they do protect smaller and lighter vehicles from sliding under bigger vehicles in the case of accidents.

INTERESTING FACTS
Undoubtedly you've seen politically incorrect bumper stickers, but have you seen the country bumper stickers? From *A* for Austria to *Z* for Zimbabwe, nearly every country has a code. Many are easy to figure out. Not so for St. Lucia, whose code is WL. That stands for Windward Islands, Lucia. If you see one with SMOM, that represents the Sovereign Military Order of Malta. EAK is on cars from Kenya—East Africa Kenya. Switzerland uses CH for Confœderatio Helvetica. And, if you see a sticker with BS, its not making any political or social statements; the car is from the Bahamas.

Convertible Top

BEHAVIOR

Opens and closes to expose the driver and passenger to the sun and wind and envy of other drivers.

HABITAT

Convertibles are found on sports cars and some sedans. Found more often in warm climates, convertibles are sometimes sported even in colder regions.

HOW IT WORKS

Convertibles can be either soft tops or hard tops. Soft tops have internal structures made of plastic and metal that support the plastic and fabric top. A motor lowers and raises the top from a compartment in front of the trunk. The rigid supports pivot and fold together in a marvel of mechanical engineering. Fully extended, it clamps to the top of the windshield to hold it in place. Soft tops usually have clear plastic rear windows that fold with the rest of the top. When lowered, soft tops are covered with a protective cloth fabric that clips in place behind the rear seats.

Hard top convertibles can be removable or retractable. Retractable tops store themselves automatically inside the trunk area. To remove the top, the driver pushes a button that activates the motor. The trunk or a separate storage area opens behind the rear seat. The windows in the doors automatically open to get out of the way and the top folds into two or more pieces as it is withdrawn to the rear. Once inside the storage compartment, the lid shuts.

INTERESTING FACTS

At the dawn of the age of automobiles, cars had soft tops or no tops. Manufacturers based car designs on horse-drawn wagons and buggies, so they made cars with similar tops. At the time, driving a car was not a practical means of transportation, as roads were poorly suited for fast driving and service stations were scattered at best. Cars were toys for the wealthy who would drive them in nice weather when a top wasn't required.

The first hard tops came out in 1910. As cars became less expensive to own and more practical to use, hard tops dominated the market. Hard tops not only shield the passengers from the elements, they also add rigidity to the car body and improve the aerodynamics by cutting drag.

Since convertibles need room to store the top when it isn't up, trunk space is usually compromised. On the next warm summer day, tell yourself that's why you don't own a convertible.

Headlights

BEHAVIOR

They light up your life—or at least the highway in front of you. Neither rain nor snow nor dark of night can stop them from illuminating the way. However, a dense fog can really cut into their effectiveness.

HABITAT

Draw a picture of an animated car driving toward you and the headlights are where you would put the eyes of the car. One is mounted on each side of the front of the car, outboard of and below the hood.

HOW IT WORKS

Most cars have halogen lights. Like traditional incandescent light bulbs found at home, halogen bulbs have tungsten filaments. The bulb itself is much smaller than an incandescent bulb and is made of quartz, not glass, and is filled with halogen gas. The halogen interacts with the tungsten to redeposit tungsten back onto the filament so it lasts longer than tungsten filaments in bulbs at home. As hot as a bulb at home gets, the halogen bulb gets much hotter—too hot to use glass, thus requiring the quartz bulb.

The silver-colored material in the headlight reflects light outward so more of the generated light is useful. Dual-beam headlights have two filaments in each headlight. Pulling and holding the high-beam lever can turn on both filaments at once.

On cars sold in the United States, low beams consume 45 watts of electric power and high beams consume 65 watts.

Some cars come equipped with High Intensity Discharge (HID) headlights that cast a blue tint. They operate like the mercury vapor lamps used in some street lighting, except that they don't have the slow start up that mercury vapor has. The gas inside (xenon) is exposed to a very high voltage electric arc that excites the gas atoms into a higher energy state. When they return to their normal state they emit photons of light. As the bulb heats up, the gas inside becomes a plasma—ionized gas. HID lights give off more light per unit of electric energy consumed than traditional headlights do.

Most cars today have sealed beam headlights. These are enclosed to prevent air from moving in or out. Each unit has a filament, reflector, and lens.

Headlights on luxury cars have cleaning or wiping systems. Mercedes has a squirter that emerges from behind a panel when you press the button to clean the lights. Other cars have mini-wipers to swish away the dirt and snow.

INTERESTING FACTS
Early automobiles relied on carbide or acetylene gas lamps. Calcium carbide mixed with water generates acetylene (C_2H_2) which burns when ignited. It also explodes and is used in carbide cannons.

Headlights in most cars are designed for use only on one side of the road. The lights are pointed downward to the outside of the road so they don't shine into the eyes of oncoming motorists.

Some cars have yellow fog lights that are better able to penetrate fog. But they are only better when the fog droplets are smaller than 0.2 microns. In most fog, yellow lights are no better than blue fog lights.

Heating Plug

BEHAVIOR

These plugs provide electrical power to block heaters and interior heaters. They allow drivers to get electric power from an external supply at home or in parking lots.

HABITAT

Some plugs hang down from beneath the front grill, waiting to be inserted into an electrical outlet. Others are built into the car fender.

HOW IT WORKS

In cold weather, engines start with greater difficulty and operate at lower efficiency until warmed up. The fuel doesn't vaporize as easily when it is cold, so the car initially exhausts more unburned fuel, adding to air pollution. And pistons are shaped so they work optimally when heated, which means that they don't fit the cylinders optimally when they are cold. This results in a further loss of energy.

Engine oil is more viscous in cold weather, which makes it more difficult for engine parts to move. And the chemical reaction in the battery that converts stored chemical energy into electricity needed to power the starter occurs more slowly. To prevent all of these inefficiencies, drivers use engine heaters.

When parking in cold climates drivers plug their cars into a source of electricity. In some places these sources have timers that either cycle on and off (to save power) or that a driver sets so the car is heated before the intended departure.

The plugs connect to block heaters under the hood. The heater warms up the engine and helps it start. It also cuts air pollution by making the engine operate more efficiently when first starting. The heaters are often inserted into "freeze plugs" in the engine block. These are expansion holes in the block so the engine can better withstand expansion of liquids during extremely cold weather. A variety of alternative heating systems are available.

Heaters for warming the inside of the car can also be connected to the heating plug. These can either sit on the floor of the car or be mounted inside the car.

INTERESTING FACTS

Andrew Freeman invented the block heater in 1946. His device (Patent #2487326) was a heating element inside a bolt. The heating bolt could replace one of the head bolts in the engine so it didn't require any other modifications to the engine.

Hubcaps and Spinners

BEHAVIOR
Like fashion models, hubcaps and spinners sashay around at incredible speeds to look good. Spinners are kinetic devices that are free to rotate even when the wheel has stopped.

HABITAT
Hubcaps and spinners cover the center of a car's wheel. Spinners and elaborate hubcaps cover the wheels of expensive cars and cars whose owners need to feel special.

HOW IT WORKS
Hubcaps are fixed decorative devices that attach to the wheel. Spinners attach to the wheel but are free to rotate. They pick up angular rotation from the spinning wheel. As the wheel rotates, friction between the bearings and the housing that holds them transfers some of the spinning energy from the wheel to the decorative spinner. When the car (and wheel) stops, the spinners continue to spin due to their angular momentum.

Other spinners are geared so they stay in place while the wheel rotates. This allows the car's logo or name on the spinner to remain upright and readable while the car is in motion.

INTERESTING FACTS
Independently rotating spinners are relatively new. David Fowlkes got a patent (#6,554, 370) for spinners in 2001.

License Plate

BEHAVIOR

They identify vehicles for law enforcement. They make the connection between owner and vehicle so you pay parking fines.

HABITAT

In the United States and most other countries, license plates are required to be viewable from the rear. Most are mounted (bolted) to the rear bumper or to the tailgate or trunk lid.

In some states, front licenses are also required. These are bolted to the front bumper.

HOW IT WORKS

License plates identify the registered owner. Each state creates its own coding system for licenses and records vehicle information numbers and other data along with the license plate numbers.

INTERESTING FACTS

License plates have been used from the very dawn of the automobile age. New York was the first state to require their use.

In the United States the standard size for a plate is 12 by 6 inches. Most license plates are made by prison inmates. License plates are punched out of mile-long coils of 0.027 inch-thick aluminum. The aluminum has to be washed and flattened. Plates are not painted; graphic sheets are glued to the plate. The corners are rounded and holes are punched for mounting. The raised numbers and letters that are unique to each plate are stamped into the plate. Prison inmates stamp each plate individually. Then the raised numbers and letters are inked. After that the plates are loaded into an oven to set the ink and adhesive (that holds the graphics).

Spoiler

BEHAVIOR

On passenger cars and trucks, the chief purpose of having a spoiler is to make the car look cooler. In race cars spoilers (wings) push the rear of the car downward to increase the traction to improve both acceleration and braking.

HABITAT

Spoilers generally are found on the rear of the car body. However, some cars—NASCAR race cars, for example—and trucks have spoilers on their roofs. Less noticeable are spoilers beneath the front of passenger cars.

HOW IT WORKS

The word *spoilers* comes from the idea that the structures disrupt or spoil the natural flow of air over the car. Technically spoilers and wings are different, although they are lumped together here.

Wings are aerodynamic devices whose purpose is to move air. In airplanes they push air downward so the plane has lift. In cars, wings are upside down so they push the car down to give it better traction.

Race cars have them over the rear or driving wheels to provide better traction.

NASCAR cars now have safety wings (called spoilers) on the roof to provide downward force when the car is moving backward at high speed. You might wonder why race cars need downward force when moving opposite to the normal direction of driving. NASCAR cars have a tendency to fly when traveling backward after a collision or spin out. A car moving backward at high speeds generates so much lift that it lifts off the ground, making it uncontrollable. Roof spoilers apply downward force to reduce the chance of lift-offs after accidents.

Passenger cars use devices to cover parts of the car to make them more aerodynamic, reducing drag. A belly pan under a car can smooth the air flow and keep it away from uneven surfaces.

Trucks use spoilers to divert air up and over their trailers. The flat front surface of a trailer presents a large drag surface. The spoiler pushes air up and over this surface.

Stylistic spoilers don't have aerodynamic or wing shapes. They sit on the rear of red sporty cars looking cool.

Windshield

BEHAVIOR
Windshields block debris and water from the interior of the car while allowing the visibility needed for safe driving.

HABITAT
Windshields occupy the space between roof and hood on the front of the car and between the roof and trunk along the rear.

HOW IT WORKS
Windshields are a sandwich of polyvinyl butyrate (PVB) between two layers of glass. The PVB holds the two layers of glass together without distorting or limiting the optical qualities. This laminate makes the windshield almost shatterproof, so if it's damaged it won't launch shards of glass into people. The windshield is glued into the window frame. On motorcycles, the windshield is often made of acrylic plastic instead of glass.

Windshield glass transmits nearly all visible light and most infrared light while reflecting most ultraviolet light. Thus, you can see out (and in, unless the windows are tinted) and the car heats up when left in the sun. However, you won't get a sunburn (which is caused by ultraviolet rays) from sunlight passing through the windshield.

INTERESTING FACTS
In the United States about 13 million windshields are replaced each year.

Windshield Wipers

BEHAVIOR

They wipe back and forth to swish the rain and snow off a viewing quadrant so you can see the road ahead. Their melodic "wipe wipe wipe" can lull you to sleep.

HABITAT

Wipers rest (park) at the base of the windshield, like a faithful dog resting at your feet.

HOW IT WORKS

Wipers are powered by an electric motor that turns a worm gear. A worm gear is a spiral of a raised edge wrapped around a metal cylinder, much like a metal screw. Worm gears are fundamentally different from other gears in several ways. They can radically increase the turning power, or torque, which is useful in applications such as windshield wipers where torque is need to push the long wiper across the windshield.

Also, worm gears can change the direction of rotation. In the windshield wiper the worm gear changes the direction of the motor shaft's rotation 90 degrees. The worm gear drives a second gear which is connected to a cam or crank. The cam or crank converts the rotary motion of the motor into the back-and-forth motion of the wipers.

Hand-operated rubber wipers were introduced in 1917 by the company that later became Trico, which is today the largest maker of windshield wipers. After World War I, the company introduced wipers powered not by electric motors, but by the vacuum pressure created in the intake manifold of the engine. This arrangement meant that the speed of the wipers was tied directly to the speed of the engine. Electric motor wipers were introduced in 1926.

Wipers for rear windows were added in 1959. Intermittent wipers were invented by engineer Robert Kearns (later sparking a series of lengthy patent infringement suits) and were introduced in 1969 on Ford's model Mercury.

INTERESTING FACTS

Windshield wipers were invented before windshields were common in cars. In their early years, cars were a fair weather mode of transportation and had no use for wipers. Mary Anderson invented a simple wiper for streetcar windows in 1903. Anderson lived most of her life in Birmingham, Alabama. But on a trip to New York she noticed how hard it was for a trolley driver to see through the windshield during a rainstorm, and this experience prompted her to invent the wiper.

Wipers for headlights were added by Saab in 1970. Now rain-sensing wipers have appeared on luxury cars.

Wing Mirror

BEHAVIOR

Wing mirrors allow drivers to see behind them along both sides of the car. Overtaking cars can be difficult to see when positioned in the driver's blind spot. Wing mirrors help drivers see places that the rearview mirror doesn't show.

HABITAT

They are attached to each of the front doors, near the forward edge.

HOW IT WORKS

Do you remember the scene in the movie *Jurassic Park* when the Tyrannosaurus rex was chasing the car? As the driver (Sam Neill) glanced back at the fast gaining T-rex, you could read "Objects in mirror are closer than they appear." As if the T-rex wasn't close enough!

Wing mirrors are not flat. They have a convex shape to capture images of a wider area. A consequence is that the images they reflect appear to be farther away than they actually are.

Because drivers adjust the seat position to fit their bodies, wing mirrors have to be adjusted so they reflect light to the drivers' eyes. Less expensive mechanisms for adjusting the mirrors include direct mechanical adjustment either with an inside lever or by pushing on the

mirror itself. Many cars allow drivers to adjust mirrors on either side with switches that control motors that rotate the mirrors.

INTERESTING FACTS
On some cars, the wing mirrors also carry turn indicators. The turn indicators are composed of LED, or light emitting diodes, set in the shape of an arrow to indicate that the car is intending to turn. Depressing or lifting the turn indicator immediately starts the LEDs blinking. The driver can barely see them, as the LEDs are positioned behind the glass in the wing mirror. However, drivers in following cars have a better angle and can see the arrows clearly.

INSIDE THE CAR

SLIDE INTO THE DRIVER'S SEAT of a new car and your eyes are drawn to the instrument panel. So many knobs, levers, and gauges grab your attention that it might be hard to focus on the road. Displays tell you the outside and inside temperatures, the direction in which you are traveling, the engine's temperature, and the radio program blasting out of the speakers. You can open rear windows, turn on seat warmers, and start the GPS. If you like feeling that you're in control, this is the seat to be in.

Air Bag

BEHAVIOR

Hides quietly in the car until provoked by a collision. Then it instantly self-inflates, coming to the rescue between passengers and injury-producing hard parts of the car.

HABITAT

Sits behind covers in the steering wheel and in the dashboard in front of the front-seat passenger. Some cars have additional air bags inside the doors. Wherever you see the letters SRS (Supplementary Restraint System) an airbag hides beneath.

HOW IT WORKS

Accelerometers, devices that detect sudden changes in speed, switch on a gas generator that is housed in the engine compartment. The fast-expanding gas flows through tubes into nylon bags that quickly inflate. Small holes in the bags allow the gas to escape moments after the collision.

The accelerometers activate the system when they detect a crash at speeds greater than 15 to 25 mph. The accelerometers are tiny electro-mechanical devices that send an electric signal. The accelerometer signals a microprocessor that in turn sends a high-current electric pulse to a heating element called a squib. The squib heats solid pro-pellants and that causes an exothermic (or heat-generating) reaction combining two chemicals to form nitrogen gas. As the bag expands, it pushes its cover away to escape its confinement at 200 mph. Less than 1/20 of a second after the car is in a collision, the air bag deploys.

The air bag itself can cause injuries when it quickly inflates and hits the passengers. People get scratched by the bag and their glasses can smash into their faces. However, air bags provide cushioning and more gradually slow the forward momentum, reducing the possibility of serious injury. As the occupant's momentum pushes on the deployed air bag, the gas is forced out of the bag through small holes. The slowly deflating air bag provides the cushioning that protects people. Some safety systems use accelerometers to activate tensioning of seat belts during a collision in addition to deploying the air bags.

INTERESTING FACTS

Since 1973 the car-buying public has been able to get additional safety by purchasing cars with air bags. In 1998 dual front seat air bags became mandatory equipment on new cars sold in the United States.

Air Conditioning

Not much value in Alaska, air conditioning makes everyone elsewhere more comfortable on a hot, sunny day. It cools and reduces humidity in the car.

The controls are integrated into the heater controls in the center of the dashboard. The working parts are under the hood.

Air conditioners work similar to the way refrigerators work. Heat is removed from air by blowing it past cooling coils. Fluid inside the coils is circulated through two parts of the system. Where the fluid evaporates in the tubes, it cools and can absorb heat from the surrounding air. Where the fluid condenses back into a liquid state, it heats up and then gives off some of its heat to the surrounding air. The cooling side of the system is located in the passenger compartment and the heating side is located under the hood.

A compressor increases the pressure of the fluid in the gas phase, which increases its temperature. The gas passes through a set of coils of a heat exchanger or radiator where it cools by contact with the surrounding air under the hood of your car. Now cooler, it condenses back into a liquid phase and passes through a pressure reduction valve, cooling further.

The liquid moves through the second set of coils where it removes heat from air blown past the coils. The heat that the fluid picks up makes it evaporate back into a gas phase and then it moves toward the compressor again. The compressor is powered by a rubber belt that is turned by a pulley on the crankshaft.

The fluid used in car air conditioning systems used to be Freon. But with the discovery that Freon released into the atmosphere depletes the ozone layer, other chemicals have been substituted. Haloalkane refrigerants are now used.

INTERESTING FACTS

Air conditioners for cars were invented about the same time as Willis Carrier invented the first modern air conditioner. Early car models were more primitive than Carrier's device. The concept for the modern air conditioner dates back to Michael Faraday in 1820.

Automatic Windshield Wipers

BEHAVIOR

Like magic they know when the windshield is wet and needs to be wiped dry. And they know how fast to swish the wipers depending on how heavy the rainfall is.

HABITAT

The sensor is most often located directly in front of the rearview mirror, on the inside of the windshield in the center.

HOW IT WORKS

Looking at the sensor you can guess the general operation. Since it is on the inside of the windshield and has no apparent holes in the windshield, it must use light. But how does it use light to detect rain?

The light it uses is in the infrared band, so you don't see it. The sensor sends out pulses of infrared light at a sharp angle to the glass (about 45 degrees).

Water on the glass changes how the light behaves at the outer edge of the glass. When the glass is dry, much of the light from the infrared source is reflected back toward the sensor. But when the glass is wet much of the light is scattered in different directions and less returns to the sensor. This effect is caused by the difference in the index of refraction (how much light bends) for air and water. Water on the windshield allows more of the infrared light to escape. When the sensor detects less infrared light returning, it turns on the wipers.

The device senses how quickly water builds up on the windshield and sets the rate of wiping accordingly. If the windshield is wet, the sensor speeds up the wiping. If it is not as wet (more light returns to the sensor), it slows the wiper.

INTERESTING FACTS

Rain-sensing windshield wipers were offered on luxury car models starting about ten years ago. Today, several manufacturers install them. The one shown here is on a Peugeot.

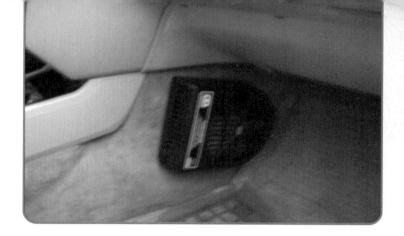

Auxiliary Heater

BEHAVIOR
Warms up the interior of your car before your scheduled trip.

HABITAT
The heater sits on the floor of your car, on the passenger side in the photo shown. It connects to an external heater.

HOW IT WORKS
This is a hedonistic addition to an engine warmer. In cold climates drivers plug their cars into a heating station (see page 26). Some heaters have timers so you can program them to start heating the car an hour or so before you start your journey.

At the programmed time electric current starts flowing into the heater. The primary heater warms up the engine. The auxiliary heater, shown here, warms up the interior of the car.

INTERESTING FACTS
Upon jumping into a car in Kiruna, Sweden, 90 miles north of the Arctic Circle, I had my introduction to this heater. With temperatures about −20° C while traveling through Lapland, this heater was a wonderful appliance to have.

Brake Light

Lights up briefly every time you start the car, when the emergency brake is set, and when the brake system has failed.

The brake light is located in the display area directly in front of the driver.

HOW IT WORKS

The brake system of a car is divided into two parts. For cars that have front disc brakes and rear drum brakes, the two different systems constitute the two parts. If one of those develops a leak and loss of hydraulic pressure, the pressure differential switch will move and cause the brake light to illuminate on the dashboard.

The differential switch is housed with the metering and proportional valves that compensate for differences between the two types of brakes. The differential switch is a piston exposed to pressures from the two different systems. If pressures remain equal, the piston stays centered in its housing. But a loss of hydraulic pressure pulls the piston to one side, which switches on the brake light. There are three colocated switches, which collectively are called the combination valve.

If the brake light is on when the emergency brake is not engaged, check the brake fluid level as soon as possible. If that isn't the problem, or if the brake light comes on again soon after adding brake fluid, have the brakes checked soon.

Brake Pedal

BEHAVIOR

A gentle push on the pedal slows the car. A sudden "Oh my gosh, I didn't see that truck!" push provides deceleration at an uncomfortable level.

HABITAT

The brake pedal is located near the floor of the driver's side, adjacent to and left of the accelerator.

HOW IT WORKS

Braking occurs when brake pads are forced to rub against spinning wheels. This action converts the kinetic energy of the car's motion into heat through friction.

Cars use hydraulic braking systems. Pressing on the brake pedal pumps brake fluid in the master brake cylinder. Most cars have power brakes that use an additional source of power, a booster, to increase the pressure in the brake lines.

Pressurized brake lines force the brake pads to rub against the rotor or drum. In disc brakes, commonly found on the front wheels, the pads are held in calipers that pinch the rotor (which is bolted to the wheel). You can see these brakes behind or through openings in the wheel. Because they are open to the air, they dissipate heat quickly. This is one of the advantages of disc brakes. Brakes can heat up to several hundred degrees in vigorous stopping and disc brakes cool faster than drum brakes, which are enclosed.

In drum brakes, often found on the rear wheels, the braking action occurs inside the wheel drum. A curved brake shoe pushes outward against the brake drum to slow the car. Springs pull the shoe back when the brake pedal is released. The brake drum is bolted to the wheel. The heat generated by drum brakes takes longer to dissipate since the braking action is enclosed inside the drum. On descents down long hills, drum brakes can loose effectiveness as heat builds up. This is why disc brakes are preferred.

Brake pads, called shoes for drum brakes and disc pads for disc brakes, wear out and need periodic replacement. Even if the pads wear out, the brakes will work. However, applying brakes will cause the sound of metal on metal rubbing and it will damage the rotor or drum.

Why use disc brakes on front and drum brakes on the rear? Cost. Disc brakes cost more. Since the front wheels supply 60 to 90 percent of a car's stopping power, it's important to have the better brakes—disc brakes—there. And it's cost effective to use drum brakes on the rear for most cars. However, sports and luxury cars don't skimp; they have disc brakes on all four wheels.

INTERESTING FACTS
Brake fluid can be one of several different types of hydraulic fluid. In the United States these are described as Dot 3 (most common), Dot 4, or Dot 5. These are made of synthetic oils, silicon, or mineral oils.

CD Player

Rapidly rotates compact discs so you can enjoy the encoded sounds.

The players are usually mounted in the center of the dashboard, adjacent to the radio.

A laser reads the CD by bouncing a beam of light off the disc. The laser light is generated by a diode laser inside the CD player. The reflected light mimics the coded marks on the CD and can be read by an electronic component called a photo diode.

The CD spins at a variable rate, depending on where the laser is reading. Reading at the start (center) of a CD, the CD spins at 500 revolutions per minute. As the song progresses, the laser is carried farther outward, toward the edge of the CD and the CD spins at a slower rate. If you look inside a CD player you can see the mechanism that carries the laser (look for a lens) in and out.

The song or other audio signal is imprinted into the CD in the form of pits or bumps. The CD is engineered to give maximum discrimination between a pit and land (area with no pits), so the photo diodes can read the encoding. The signal of digital bits is fed through the electronics, is amplified, and sent to a headphone or speaker.

CD inventor, James T. Russell, had a difficult time finding any companies to license his invention for storing data on a plastic disc. He had been motivated to find a better way to record music with a system that didn't have to have mechanical contact with the record. He worked for several years before winning a patent in 1970. Sony was one of the first companies to realize the possibilities of the technology and purchased a license. Russell has kept on inventing, winning more than two dozen other patents.

Child Car Seat

BEHAVIOR
Provides a safe and comfortable place for children to ride.

HABITAT
Found in the rear seats of all cars with the yellow window sign that announces "Baby On Board." They can be found in many other cars as well.

HOW IT WORKS
Padded belts hold the child securely in the seat. Facing to the rear of the car allows the force of sudden stops to be distributed evenly over the child's back, which is much less dangerous than forward-facing belt systems where the force is concentrated where the restraining belts are.

The security of the child seat depends on how it is affixed to the car. Usually adult seat belts are clipped around the base of the child seat. A 2002 regulation now requires that cars come equipped with an anchor behind the rear seat to which the top of the child's seat can be attached.

When children outgrow the child seats, they should use the adult seat belts while sitting on booster seats. The seats raise them high enough so the adult belts contact their bodies at safe positions.

INTERESTING FACTS
Not a new invention, child car seats were first introduced in the 1920s. These first seats were not a safety device for the child but were designed to restrain the child from crawling around the car and causing problems for the driver. Later models weren't much more effective for safety until the National Highway Traffic Safety Administration ordered improvements in 1971. Companies had conducted research to improve designs and the testing was improved greatly when child-size crash dummies replaced dolls in the tests.

Tennessee accelerated the use of child seat belts by mandating their use in 1978. Within ten years the other states had enacted similar laws.

Cruise Control

BEHAVIOR

On long rides, especially on interstate highways, cruise control provides a welcome relief for your right foot. Instead of holding your foot in a rigid position for mile after mile ("Exactly how wide is Kansas?"), you set the cruise control for the speed you want and pull your foot off the accelerator.

HABITAT

The buttons or switches for cruise control are most often found on the center panel of the steering wheel. Some cars have them on the turn indicator stalk or a similar stalk that is just for the cruise control.

The control system can be seen under the hood. Look for a distinctive pair of cables running over pulleys, side by side. One of these is the throttle control connected to the accelerator pedal and the other is a throttle control operated by the cruise control system.

HOW IT WORKS

The cruise control system controls the speed of the car by adjusting the throttle, just as you do with your foot on the accelerator. However, it uses a motor to adjust the throttle position. Both controls, from the accelerator pedal and from the cruise control system, connect to the throttle so you can see and feel the accelerator pedal move as the cruise control adjusts the throttle.

The motor that moves the cable that adjusts the throttle position only moves back and forth. It can be either an electric linear actuator

or a vacuum actuator. The vacuum actuator uses the vacuum pressure from the engine to move a diaphragm back and forth in response to electronic controls. Electric actuators are solenoids or electromagnets that pull a metal core back and forth as the electric current changes.

A microprocessor runs the cruise control system. It gets input from several sensors and switches, and sends a control signal to the actuator. The driver-operated controls provide one set of input data. The brake pedal and clutch pedal (for manual transmission cars) have sensors that tell the microprocessor to disengage the cruise control when either pedal is touched. The microprocessor also gets data on the car's speed (from one of several sensors) and on the position of the throttle sensor.

New cruise control systems that automatically keep your car a safe distance from the car ahead have been developed. A light (laser) or radio (radar) beam determines the distance to the next car. The system can also alert drivers to an imminent collision when cars get too close. These systems can automatically apply brakes and tighten the seat belts when it senses an imminent collision.

INTERESTING FACTS
The cruise control was invented by a Ralph Teetor, a blind inventor who was annoyed by his lawyer's driving. The lawyer would slow down and speed up depending on whether he was talking or listening. In addition to Teetor's several inventions, he ran a car parts manufacturing business.

Defrost System Control

BEHAVIOR

With a flick of your wrist you send a stream of warm, dry air over the inside of the windshield to remove the condensation and improve visibility of the road ahead.

HABITAT

The defrost controls are part of the heating and air conditioning system, usually located in the center of the dashboard. The vents for the defrost system are aligned under the windshield. Rear windshields have de-ice/defrost wires embedded in them.

HOW IT WORKS

Warm air is blown across the windshield to evaporate the condensate that fogs your vision. Warm air holds more moisture than cold air so air is warmed before being blown out the vents.

However, before the air is warmed, it is *cooled*. Does that make sense? Cold air can hold less moisture than warm air, so cooling the air first through the air conditioning system reduces the amount of water in the air. Cooling reduces the humidity before the air is warmed so it emerges with very low vapor content and a high capacity to absorb water vapor. Air is forced out of vents so it moves along the interior side of the windshield and the driver's and front passenger's windows.

Heating wire embedded between sheets of glass in the rear windshield clears the fog there. Electrical current passing through the wires heats them, which melts snow and ice on the outside and clears away any interior fog. These defrost systems are operated by a separate control and shut off automatically.

DVD Player

DVDs are used to entertain those sitting in the rear seats. They can watch their favorite cartoons or other movies to make the miles fly by.

HABITAT
DVD player controls can be located either in the center of the dashboard or between the front and second row of seats in a minivan. The screen drops down from the ceiling, secured on a hinge, so only those in the back seats can view it.

HOW IT WORKS
The Digital Video Disc (DVD) works like a compact disc (CD), but it can store about seven times as much data. A DVD has enough data capacity to store a full-length movie, which is why they are so popular in cars.

Both DVD and CD players use lasers to read the data. Both discs are 120 mm in diameter and made of polycarbonate plastic with a thin metal coating. However, DVDs are somewhat slimmer and that allows them to use a different lens that can discriminate more densely packed data.

A DVD player uses a laser operating at 650 nm (nanometers, or one billionth of a meter), while a CD player operates at 780 nm. The shorter wavelength for DVD players allows the laser to read finer detail. DVDs have more data per square inch of surface than CDs, and less data storage is allocated to error correction. The various changes from CD format allow the DVD to hold so much more data.

INTERESTING FACTS
Like CDs, DVDs are read from the inside (near the center) toward the outer edge in a long spiral. The spiral for a DVD is over 7 miles (11.3 km) long.

Flares (Fusee)

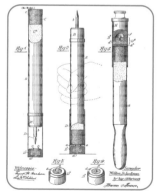

Hopefully they sit unused in the trunk of your car. When needed and ignited, they emit a blindingly bright light to warn other drivers to avoid your disabled car alongside the road.

HABITAT

Usually they are found in a plastic package in the trunk or other storage compartment of a car.

HOW IT WORKS

Flares or fusees are containers of chemicals that, when ignited, combust in a heat-generating chemical reaction. Shaped and sized like a stick of dynamite, a fusee is made of a waxed cardboard cylinder. The chemicals inside flares include fuel (charcoal or other burnable material) plus a pyrotechnic composition that releases energy once ignited.

Flares typically burn for 15, 20, or 30 minutes. Most are designed to be laid on the roadway, but some have spikes so you can set them vertically along the shoulder of the road.

To ignite a fusee or flare you rub two surfaces together, similar to lighting a match. First remove the top cap that protects the scratch surface. Then twist to remove the second cap and thus expose the fusee igniter button. Rub the scratch surface (on the cap) across the igniter button. This, by the way, is *not* the time to think about where you are going to put the flare. You want to have that figured out well before igniting it.

Wilton Jackson invented the strike-to-start fusee in 1899.

Four-Wheel-Drive Shifter

BEHAVIOR

The shifter allows drivers to switch into and out of four-wheel drive. A four-wheel drive system provides additional traction that is needed for driving off the road or on icy roads.

Having four-wheel drive makes you the envy of your neighborhood, useful in itself even if you never need the additional traction.

HABITAT

Controls for four-wheel drive are found adjacent to the gear shift levers next to the driver. They are often mounted in the floor between the driver and front seat passenger.

HOW IT WORKS

There are many variations of four-wheel drive systems, including full-time, All Wheel Drive (AWD), or four-wheel drive (4WD). AWD is meant to be used on roads in normal driving as well as for low-traction situations. Vehicles with 4WD are switched from two-wheel drive into four-wheel drive when encountering low traction.

The aim of these systems is to provide power to all the wheels that have traction. Having four wheels push the car is better than having just two push when there is the chance that the tires might slip.

The problem with engineering four-wheel-drive systems is getting the right amount of torque or tuning power to each wheel and allowing each wheel to spin at the right speed. Turning a corner requires that each wheel travel a different distance at a different speed. Locking all the wheels together so they turn at the same rate will cause the tires to drag on the ground. This makes for a rough ride and puts excessive wear on the tires.

Traditional 4WD systems require the driver to shift from normal two-wheel drive into four. Older models forced the driver to stop the car, get out, and "lock the hubs" on the front wheels, and then shift into four-wheel drive. This had obvious drawbacks, since the driver had to anticipate where he needed the 4WD and then get out to lock the hubs in what often were the worst environmental conditions.

Systems available today are easily confused. AWD systems provide power to all the wheels at different rates. Each wheel gets the power it needs. Full-time 4WD can be locked so all the wheels receive the same power or unlocked for highway driving, allowing the front and rear wheels to turn at different speeds. A 4WD car can be switched between two-wheel drive and four by the driver.

The transmission provides power from the engine to a transfer case on a part-time 4WD system. The transfer case distributes power to the second axle, splitting the power between front and rear axles. Locking the hubs on the wheels on an axle connects the wheels to the power provided through the differential. This forces the wheels to spin at the same rate as the wheels on the other axle—great when you need full traction but bad on dry pavement.

In low traction and off-road situations, one or more wheels may loose traction or even lift off the road. To solve this problem, a system of traction control applies brakes to the wheel that has lost traction. This prevents the free wheel from spinning wildly and distributes more torque to the other wheel.

INTERESTING FACTS

Like much of the early history of cars, electric vehicles led the way with four-wheel drive. Ferdinand Porsche demonstrated a car that had four electric motors, one mounted in each wheel, in 1899. The first gasoline cars with four-wheel drive were built for racing or as sports cars. World War I saw the use of 4WD trucks, but they were not used widely until World War II. Then the Jeep became an American icon. Jeep, and later Land Rover, produced 4WD vehicles for the public. Today, with new systems for sharing power between wheels and for providing braking power to wheels, four-wheel drive has become a popular option.

Fuel Gauge

Unwatched for many hours on end, it becomes the bearer of bad tidings when gas stations are miles away. It lets you know how much gasoline or diesel fuel remains in the tank.

HABITAT
The fuel gauge occupies high-value real estate on the dashboard, directly in front of the driver.

HOW IT WORKS
The gauge you see is but one part of the system. The position of the needle directly represents how much electrical current is flowing through the circuit that feeds it. The gauge is wired to a sensor in the gas tank. Inside the dark recess of the fuel tank a float rides atop the fuel. As the fuel level drops, the float drops with it. Connected to the float is a variable resistor that controls the current in the electric circuit that is monitored by the gauge.

After you have taken out a second mortgage to pay for a tank of gas, the float will ride at its highest position and the electric current flowing through the gauge circuit will be highest. This drives the needle on the gauge to the big *F*. As the fuel level drops, the float rides lower and reduces the current in the circuit. Too soon you find yourself watching the gauge needle, trying to push it back up to *F* with your eyes.

Should the float system fail, the gauge needle will point to the dreaded *E* even if you have a full tank of gas. Thus, the system should never show a higher level of fuel than exists in the tank.

Fuses

BEHAVIOR

Fuses are the safety valves in any electrical system. Should something go wrong in a circuit causing the current to rise to unsafe levels, the fuses break, interrupting the flow of electricity. Without fuses a high current could damage the wiring or expensive electrical components that the wire connects.

HABITAT

Fuses are housed in a protective fuse box. This can be inside the passenger compartment behind a plastic door adjacent to the driver's left knee, or in the engine compartment. Check the owner's manual for the location.

HOW IT WORKS

A fuse is a thin strip of metal conductor that has two electrical contacts and is encased in a plastic housing. As the current in a circuit rises, heat generated by electrical resistance rises too. Inside the fuse the conductor will melt if the current rises too far. When it melts, the circuit opens, stopping the dangerous flow of electricity.

A car's electrical system is divided into several independent circuits. Each electrical circuit is protected by a fuse. This allows different circuits to operate at different voltages and currents. For example, the starter motor draws much more current than the lights or radio need, and it has a fuse with a higher current rating.

Automotive fuses are designed for up to 24 volts and up to 30 amps, although most have lower current ratings. Most common are blade-type fuses that plug into slots in the fuse box.

Many cars come with spare fuses inserted into dummy sockets inside the fuse box. If a fuse fails, you will notice that one system or several systems won't operate. Check the owner's manual or the inside of the fuse box for a diagram that shows which fuse protects each system. Extract the fuse from the position protecting the system that isn't working and insert one of the spares. If your luck is good, the change of fuses will have solved your problem. If the new fuse fails soon after you've installed it, you know you have a more serious problem requiring the help of a mechanic.

Old style fuses were glass tubes with a thin metal wire running through them. Both ends were capped by a silver-colored metal. This style fuse is still used in some household electronic devices.

Glove Box

BEHAVIOR
It tends to contain more stuff than should fit. It is the repository for the pieces of paper that you must keep with the car, the tire pressure gauge, ice scraper, maps, key chains, vehicle registration, proof of insurance, pens, and a flashlight with batteries that have died many months ago.

HABITAT
Found in the dashboard on the far right side, in front of the passenger seat.

HOW IT WORKS
Usually lockable, especially in convertibles and Jeeps, it opens with a latch button. The door falls down to provide the front-seat passenger with a cup holder or two.

INTERESTING FACTS
The glove box was designed in early cars as a place for the driver to keep gloves. Cars were not enclosed and didn't have heaters so drivers needed gloves in cold weather, and they stored them in a compartment conveniently built into the dashboard.

Global Positioning System (GPS)

BEHAVIOR

It shows you the way to go home. Or to that meeting that you are late for. It provides key navigation information in an easy-to-follow visual format and orally and instantly updates the directions when you miss the critical turn.

HABITAT

GPS units are usually mounted on the dashboard where the driver can easily see them. Some are removable so they can be locked away for security when the car is parked.

HOW IT WORKS

Two dozen navigation satellites orbiting the earth continuously transmit time signals at microwave frequencies. Each satellite has an atomic clock so it can deliver time accurately. The receiver in a car needs to receive the time signals from at least four satellites to compute its position.

The computer inside the GPS unit computes the distance to each satellite based on the time it takes each signal to arrive. To do this the computer has to know what time it is and where the satellites are. Its first task, once it has locked onto signals from four or more satellites, is to compute the time and then the distance to each satellite. The amazing thing is that it can do this to an accuracy of only a few feet even though the satellites are thousands of miles away.

The GPS unit computes the position of the car in terms of latitude and longitude. This, of course, isn't useful to you unless you know the latitude and longitude of where you are going.

The location information is integrated onto a map database so drivers can see where their cars are in relation to street addresses, roads and bridges, and other key points. The map data is stored in a factory-installed ROM (read-only memory). However, new information and more detailed information for a particular region can be added to the GPS by downloading it from the Internet.

INTERESTING FACTS

When he was vice president of Raytheon Corporation Ivan Getting came up with the idea for GPS, and Professor Bradford Parkinson of Stanford was the chief architect and first program manager of the GPS system. Both men were inducted in the National Inventors Hall of Fame for inventing the GPS system.

GPS grew out of the U.S. Navy's efforts to provide more accurate navigation for its ships—especially its ballistic missile submarines. Until the year 2000 civilians could receive only degraded information from the GPS system, as the military didn't want to allow potential adversaries to have access to the most accurate navigation information.

Hand-Cranked Window

BEHAVIOR

You rotate the handle and the window moves up. Rotate it in the other direction and down she goes.

HABITAT

A handle is mounted on the inside of the driver's and front passenger's door, in about the middle of the door. For backseat passengers, the handle is either mounted in the door (for four-door cars) or on the sides of the car directly beneath the window. The hand-cranked window is found only in less expensive cars these days; most often the windows are operated by electric motors.

HOW IT WORKS

The handle in the door is connected to a small gear that meshes with a larger gear. This larger gear would take up lots of room inside the door, so instead of using a full, 360 degree gear, a smaller partial gear is used. It is only about one-quarter of a complete circle, and is called a quadrant. A long lever arm is attached to the quadrant and that pushes the window up and down.

The gearing—a small gear driving a much larger diameter gear or quadrant—yields a mechanical advantage, which you need to lift the heavy glass window. Two lever arms connect the quadrant to the bottom of the window pane and these provide additional leverage to lift the window.

Heater

Slow to warm, it then blasts out the heat on a cold winter morning. It keeps you toasty warm once it has time to get started. Many cars have heaters that will broil the occupants after a few miles.

HABITAT

The controls are found in the center of the dashboard. The heater itself is mounted beneath the dashboard. Vents are mounted strategically in the dashboard, sometimes in the driver's and passengers doors, and infrequently in the rear of the car to warm up the shivering backseaters.

HOW IT WORKS

Car heaters use the engine's warmth to warm up the passengers. A small radiator, called a heater core, is housed inside the dashboard. Hot water from the engine circulates through it, pressurized by the water pump. Heater hoses, which are smaller than the hoses that carry coolant to and from the radiator, deliver water to the heater core.

The heater core is made of metal pipes that sport fins to give them more surface area to transfer heat to air passing between the pipes and fins. Thermostat controls direct or block the engine coolant from entering the heater core. Dual control heat systems that allow separate passenger and driver controls have, in essence, two heater cores. Fans blow air over and through the heater core, through ducting, and into the passenger cabin.

INTERESTING FACTS

The first electric heaters in vehicles were in electric streetcars. Thomas Ahearn, the Canadian Thomas Edison, founded the Ottawa Electric Railway Company in 1891. To warm passengers in the cold Ottawa winters, he had electric heaters installed in the cars. In addition to his many other innovations and inventions, Ahearn was the inventor of the electric stove and the first person to cook a meal using electricity.

Key Fob

BEHAVIOR

Allows you to lock and unlock the car while dozens of feet away in the parking lot. Most units also have panic button alarms and trunk releases. Some cars have keyless ignition switches, too.

HABITAT

It hides out in a pocket or pocketbook until you pull it out to fiddle with the buttons or insert it into the ignition.

HOW IT WORKS

These keyless systems are keys that are electronically encoded. The key fob transmits the coded signal as a 315 MHz signal (in North America). The fobs come from the factory preprogrammed with unique codes. They are paired with automobiles that "learn" the code from the fob. A computer in the car is set to receive the code of a fob and "remember" it as the key. The fob sends the identify code, which is a long string of numbers, plus the code to tell the car what you want it to do. A basic system lets you lock or unlock the doors, open the trunk, or sound an alarm. More advanced systems start the car, too.

New systems change the code every time you use the fob. This prevents someone from setting up a radio receiver in a parking lot to

capture the code and open or steal cars while you are shopping. Both fob and car have identical programs that generate new codes each time you use them. So after each use both perform the same mathematical algorithm that generates a new code. But if you start pushing the buttons on your key fob while you're a full parking lot away from your car, the fob and car will be out of sync. To prevent you from being locked out, the car's computer will accept any of the next 256 codes that the coded fob might transmit. So as long as you don't punch the button more than 256 times while out of range, the car doors will unlock.

If you have multiple fobs for your car, the car's receiver will accept codes from any of the two to four fobs. Each one acts independently with the receiver, generating and sending codes.

Of course this system isn't foolproof. If you lose one of the fobs in the cracks between the cushions of your sofa and find it months later, after using the other fob several hundred times, the recovered fob will be out of sync. But all is not lost. The car manufacturer has procedures to reprogram the car's coding system to get all the fobs working again. With some models you can reprogram the fob by turning the car's ignition on and off several times and then operating the fob.

Odometer

BEHAVIOR

It gives a permanent record of the distance a car has traveled during its lifetime and allows travelers to misjudge when they will arrive at a distant place.

HABITAT

Odometers are afforded the honor of being located dead center in front of the driver, just below the speedometer.

HOW IT WORKS

Older cars use mechanical odometers and newer cars use electronic ones. Mechanical odometers use flexible metal cables to spin the geared display system in the dashboard. The display is geared so that many revolutions of the cable make the last digit on the right, the tenths of a mile digit, rotate one full revolution. As it does, it trips the adjacent display to the next higher number. Each display engages the display to its left when it has completed a revolution.

The cable is spun by a gear attached to the transmission shaft, thus making the measurement independent of the gear used. If it were driven by the engine, a car traveling in low gear could record the same mileage as one traveling much faster and farther in high gear. Nonetheless, driving the car in reverse will reduce the mileage shown on the odometer.

Driving a car with an electronic odometer in reverse, however, won't decrease the mileage displayed; it will increase it to reflect the total miles driven. As with mechanical odometers, digital or electronic odometers are connected to the transmission shaft by a gear. A sensor, either magnetic or optical, detects each revolution of the gear, either forward or backward. The sensor sends the signal to an engine control unit.

The engine control unit records data from several other sensors and periodically sends them to the proper displays in the dashboard. The engine control unit allows communications between several sensors and displays with a single set of wires rather than having wires for each sensor and display. It also controls many of the functions of the engine. The critical part of the engine control unit is a microprocessor.

INTERESTING FACTS

Ancient Romans, 2,000 years ago, were the first to use odometers to measure distance. They designed odometers for chariots so they could measure the length of their highways. Odometers were also used in ancient China.

William Clayton invented the modern odometer in 1847 to measure how far his wagon traveled each day. He was charged with counting the number of turns a wagon wheel made throughout each day and multiplying the total by the wheels circumference to estimate how far the wagon traveled. Soon tiring of this daily grind, he designed an odometer and had a carpenter build it. He used it on a trip from Missouri to Utah—a trip with way more rotations of a wagon wheel than anyone would want to count. Curtis Veeder invented the odometer for cars in 1906 (Patent #833,355).

Parking Brake

Prevents your car from rolling down that steep San Francisco hill. And if ever your brakes fail, you can pull on the parking brake, which is why it is sometimes called an emergency brake.

HABITAT

To engage the parking brake you either pull on a handle adjacent and to the right of the driver, or push on a foot pedal to the left of the driver. The brake itself is out of view beneath the car.

HOW IT WORKS

Most cars have parking brakes only on the rear wheels. Some front-wheel drive cars have the parking brake on these wheels instead.

Parking brakes are a mechanical system, operating without the hydraulics systems used by the other brake systems. You pull up the lever or depress the pedal to set the brake and engage a pawl in a ratchet. This pulls on a steel cable that engages the brakes in the two wheels. The pawl and ratchet hold the brake on until you release them. Some cars with automatic transmission automatically release the parking brake when you move the transmission out of park.

The parking brake uses the existing brakes, but does so without the hydraulic system. So if the hydraulics fail (a hydraulic line breaks, for example), the parking brake will still work. Cars with four-wheel disc brakes have an additional drum brake that is activated by the emergency brake handle.

New models have electric parking brakes. An electric motor either tugs on the cable that engages the brake or a motor directly squeezes the brake calipers.

Power Window

BEHAVIOR

Power to the windows! Indeed, it is very nice being able to open any window from the driver's seat with the push of a button. You probably don't often want to open or close the rear windows while traveling at highway speeds, but when you do those buttons are great. However, if they fail (broken belt, switch, or motor), driving with an open driver's window on a winter day becomes a memorable ride.

HABITAT

The buttons are located either on the driver's door or on the center console.

HOW IT WORKS

Car windows are heavy. If the belt that pulls up the window breaks, you will find out how difficult it is to pull up the window by hand. Image how powerful the motor must be to raise the window. Actually, the motor has it pretty easy as it doesn't directly drive the window up and down, but instead turns a gear and lever system that transform the motor's high rotational speed into substantial torque or turning power. Manual car windows use the same type of levers to make it easy to roll up the window.

Depressing one of the window buttons or switches provides power to the motor that lifts and lowers that window. When the window gets to the top (or bottom) and can't travel any farther, the motor tries to work harder. A circuit senses the increase demand for current and shuts off the motor. Many cars have more elaborate switching systems that control the functions of several electrical devices such as door locks, seat position, and power mirrors.

Holding down the button for the driver's window for more than a brief moment lowers the window all the way. This feature is called "express down."

Radar Detector

BEHAVIOR

They squawk as you drive down the highway and occasionally bark protectively to let you know that a radar speed trap awaits you. They detect police radar, hopefully in time for you to slow down.

HABITAT

The radar sensor is mounted behind the car's grill, out of sight. The visible part of the detector is mounted inside the car on or near the dashboard.

HOW IT WORKS

The word radar comes from radio direction and range, which indicates the initial use for this technology was finding (military) targets, mostly airplanes, and figuring out where they were (direction from the radar station, and distance or range).

Radar uses electromagnetic waves in the radio frequency range. A police unit sends out a radio wave at a particular frequency. The wave reflects off solid objects and a tiny portion of the outgoing wave reflects back to the radar unit. There a radio wave detector will sense the signal.

Of course, it isn't enough for police to know that your car is there— they need to know how fast you are driving. So rather than report your presence, the radar gun measures how much the returning radio signal frequency has changed from the signal it sent. If your car is parked along the side of the road, the returning signal will have the same frequency as the outgoing signal had (assuming the officer is standing still, too). But if you are driving toward the radar gun, the returning signal will have a higher frequency and the magnitude of the change is related directly to your speed.

Imagine that one pulse of radar reflects off your car and a second later another pulse reflects off your car. By the time the second pulse reaches your car, the car's position will have changed (if you are moving). The faster the car is going, the more the position has changed. So although the two radar pulses were sent a second apart, they will be received less than a second apart since the second one traveled a shorter distance. The radar gun measures this change, called the Doppler shift.

Radar detectors are effective in that they can pick up a weak signal, one that might be too weak to reflect off the car and travel back to the radar gun. So the detector can "see" the radar gun before the radar gun can see the car. This assumes that the radar gun is on while you are approaching.

As in all technology battles, both sides constantly improve their hardware. Police now increasingly use laser detectors called LIDAR instead of radar. And speedy drivers use LIDAR detectors, or put coatings on their cars (stealth technology) to hide from LIDAR. Unlike radar, LIDAR doesn't use the Doppler shift. It measures the distance from the gun to the car and compares the distance between successive pulses. Large differences in the distance indicate high speed.

Radio

BEHAVIOR
It brings you acoustical enjoyment, football action, news, weather, and traffic advisories.

HABITAT
The radio occupies the center of the dashboard. Located not as close to the driver as the car controls are, the radio is reachable by both driver and front seat passenger. But passengers, don't touch that dial.

HOW IT WORKS
Radio stations broadcast in either AM (amplitude modulation) or FM (frequency modulation). When you select a station to listen to you tune your radio to one frequency band. The radio signal you select is vibrating at that frequency. AM stations alter the amplitude of the vibration to carry the sounds of the annoying talk show hosts. FM stations alter the frequency that is superimposed on top of the carrier frequency.

In both cases, the electric signals are amplified and then sent to an antenna. The antenna converts the pulsating electric signals into electromagnetic waves that propagate at nearly the speed of light. Wires carry it into the receiver. The antenna also receives hundreds or thousands of other electromagnetic signals from other radio stations, cell phones, and space aliens. The radio receiver filters out all the signals, leaving only the one frequency you selected. It amplifies the signal and sends it to the speakers where the pulsating electric signal is converted into the sound of Rush's voice. Turn that off!

When car radios were first sold the price was about 25 percent of the cost of a new car. Not like the compact integrated circuit models of today, these early radios were bulky and required a massive rebuild of a car's dashboard to install, requiring several days of labor. On top of these difficulties, short circuits commonly caused fires.

The first commercially available car radios were available in 1926. The manufacturers faced several technical problems including how to minimize interference from the car's electrical system, which generated radio static. Isolating the radio from the car's electrical system improved the reception, but caused other problems until Paul and Joseph Galvin invented a new electric circuit. Their circuit design not only made car radios work well, it also launched their new company, Motorola.

Sales of both home and car radios surged in 1930 when, despite the start of the Depression, people had to have them to catch the new NBC radio show *Amos 'n' Andy*. This sales growth led to more innovations that improved the quality of the radio.

Still, drivers in the 1930s had to endure problems that we don't face today. When you stopped the car, you had to turn the radio off or face the possibility of draining the car battery. Before police had their own two-way radios installed in patrol cars (which started in 1931), they broadcast police calls on public stations. This allowed officers and everyone else, including the people perpetrating crimes, to know what was going on.

FM radio came to cars in 1951, nearly twenty years after the first FM broadcasts. But the largest technological change was the introduction of transistorized radios in 1958.

Rearview Mirror

BEHAVIOR

The rearview mirror allows you to spot the nice police officer coming up behind you. You can tilt the mirror at night so headlights of cars behind you don't shine into your eyes.

HABITAT

Front and center, the rearview mirror requires prime viewing space, directly behind the windshield.

HOW IT WORKS

Mirrors are pieces of glass that are coated on one surface with a thin layer of metal. Light reflects off the "silvered" surface. Rearview mirrors are made of a wedge of glass that is silvered on the back side. Thus, unlike most mirrors, rearview mirrors have front and back surfaces that aren't parallel to each other.

During daytime driving, light from behind the driver passes through the glass and reflects off the silvered surface so the driver can see that tailgater.

This isn't the entire story, however, as a small part of the light (about 5 percent) also reflects off the front surface of the glass. The driver's eyes only see the much stronger image that reflects off the rear (silvered) surface.

At night, the driver tilts the mirror to avoid being blinded by the headlights of following cars. The light that reflects off the silvered surface now reflects down and away from the driver, and the much weaker reflection off the front surface of the glass reaches the driver's eyes.

INTERESTING FACTS

The first use of rearview mirrors as a driving aid was in the first Indianapolis 500 race in 1911. All the cars except one had a mechanic ride with the driver to tell the driver where the other race cars were. The one driver without a ride-along mechanic, Ray Harroun, borrowed an idea he had seen on a horse-drawn wagon and installed a mirror so he could see cars coming up behind him. Elmer Berger is given credit for inventing the rearview mirror, taking Harroun's idea and making it a practical appliance for cars.

Seat Belt

BEHAVIOR

Seat belts keep you from flying out through the windshield in a head-on collision. Stopping the car instantly while unrestrained would catapult the passengers forward—until they impact solid objects. Seat belts make you part of the car and distribute the forces of sudden impacts over larger areas of your body.

HABITAT

Hopefully, they cover you and all your passengers whenever you are moving in a car. When not in use, the belt is retracted into a housing along the outside of the passenger's seat at the floor.

HOW IT WORKS

Seat belts restrain your forward motion when the car suddenly slows or stops. The belt, which once extended only over the lap, now restrains both your hips and shoulder and distributes your momentum across your pelvis and rib cage. Most common among seat belts is the three-point harness that provides a lap belt and shoulder belt.

A sudden lurch to stop, perhaps when you realize that you're about to run into the car ahead, causes a centrifugal clutch to engage a locking ratchet in the seat belt. How is that done? Excess seat belt is rolled around an axle. In a sudden stop, the passengers slide forward, extending the seat belt as they go. As the axle spins quickly, weights on a clutch are thrown outward due to centrifugal force. These push a metal bar into the ratchet to stop further motion. The metal bar, or pawl, is securely attached to the car. A pawl is one of two parts of a ratchet system; it catches the ratchet gear to prevent it from turning.

Other systems replace the centrifugal clutch with a momentum clutch. Essentially, it is a weight held vertically that continues to move forward when the car suddenly stops. As it moves forward it engages a pawl that jams into a ratchet gear, stopping the belt from unwinding.

To release the belt you provide it some slack. This allows a spring to pull the clutch back to its normal position. The belt is then allowed to wind up on a spool on the shaft.

INTERESTING FACTS

Sir George Cayley invented seat belts in the 19th century. Cayley was a prolific inventor and renowned scientist, and is considered the father of aerodynamics. The first U.S. patent for a seat belt was issued to Edward J. Claghorn in 1885. Although some car companies started offering seat belts in the 1950s, 1964 saw the universal adoption of seat belts for front-seat passengers, and 1968 for rear-seat passengers. In 1973 Volvo developed an automatic seat belt that is now standard in some car models.

Speedometer

BEHAVIOR

Like a stern third-grade teacher, the speedometer keeps you from breaking rules and from having fun on the open road. It provides an estimate of your car's speed.

HABITAT

The speedometer sits directly in front of the driver. It is usually viewed through the steering wheel.

HOW IT WORKS

The system in use for nearly 100 years uses a flexible metal cable that rotates as the drive shaft rotates. A gear attached at the end of the transmission rotates and drives the metal cable. The cable runs from the transmission to the back of the odometer in the dashboard. A magnet is attached to this end and as the car moves, the magnet spins.

The spinning magnet creates electric currents (eddy currents) in a metal cup and these generate a magnetic field that moves the needle in the odometer. Opposing this movement is a spring that has been calibrated so its length is equivalent to the speed of the car. The faster the magnet spins the more current it generates and the greater the force that stretches the spring. Oops, you're over the speed limit!

Newer electronic systems eliminate the flexible cable, which can break, and instead sense the number of rotations of the drive shaft in a set period of time. As the car speeds up the number of rotations per second increases. The frequency of the rotations (revolutions per second) is carried by wires to the speedometer where it is converted into miles per hour and (usually) displayed in digital format.

INTERESTING FACTS

The speedometer was invented before cars were on the road. Josip Belušić invented it in 1888.

The first speeding ticket was given in 1896 to a driver in Great Britian. The driver was careening along at a whopping eight miles per hour . . . in a two-mile-an-hour zone.

Steering Wheel

BEHAVIOR

Allows drivers to control the direction of the car's motion. Turning the steering wheel rotates the front wheels to the left and right.

HABITAT

Steering wheels occupy the space nearest to the driver. They are located within easy and comfortable reach. In the United States, Canada, and Mexico steering wheels are located on the left side. They are mounted onto a steering column.

HOW IT WORKS

The steering column holds the steering wheel. Inside the column is the steering shaft that mechanically conveys the steering motion to the gearing that moves the wheels. The column also houses the wires connecting the various instrument switches mounted on the steering wheel and on the column. These include the ignition switch wires, horn, cruise control, turn signals, windshield wipers and washers, and sometimes the headlight control.

The steering wheel is connected to the steering shaft, which is a rod that transfers the rotations of the wheel to a steering gear. Two types of gear are common: rack and pinion and worm gear. In rack and pinion, a pinion (or drive) gear is mounted on the end of the steering column. It is fixed in place, but rotates as the steering wheel is turned. It moves a flat rack gear (a row of parallel teeth cut into a metal bar) that moves the tie rods that connect the two front wheels.

In worm gear steering, the steering shaft rotates a worm gear that moves a lever arm (Pitman arm) connected to the tie rods. A worm gear has screw-like threads cut into a shaft.

Since 1963, many cars allow drivers to adjust the angle of the steering wheel. A lever on the left side disengages a ratchet and allows the wheel to pivot on the steering column.

INTERESTING FACTS

Like boats, early cars were steered with tillers, which are simple bars that act as levers. The tiller or lever rotates a rod that rotates the front axle. Automaker Packard switched to steering wheels in 1899 and soon the other manufacturers followed.

Tachometer

The needle jumps up and down as you depress the accelerator. Rrmmm, rrmmm, and the needle moves wildly. It shows you the crankshaft's speed of rotation. The units are given in thousands of revolutions per minute. The display shows a caution zone (yellow) and a danger zone (red) that is the maximum engine speed. "Redlining" an engine is to operate it at its maximum speed. Operating the engine above the red line can cause permanent engine damage as moving parts heat up, expand, generate more friction, and eventually bind and stop moving.

HABITAT

The tachometer, or tach, sits in the middle of the instrument panel in the dashboard, immediately in front of the driver. The sensor sits next to the speedometer and is usually the same size and design as the speedometer.

HOW IT WORKS

Most tachometers are eddy-current tachometers. As the engine turns, a magnet on the crankshaft creates an electrical current in a surrounding coil. As the engine revs faster, the magnet generates a greater electrical voltage. The instrument in the dash panel displays the voltage generated. However, rather than show the voltage per se, it is calibrated to show the speed of rotation that generated the voltage.

The tachometer and speedometer both measure rotations of engine-driven shafts. But the tachometer measures the engine speed, or speed before the transmission, and the speedometer measures the drive shaft, or after the transmission, speed. As the car accelerates from a stop, both speed and engine revolutions rise until you (or the automatic transmission) shift gears. Then the engine speed drops momentarily while the car speed continues to increase.

Temperature Gauge

BEHAVIOR

Usually the needle of the gauge hovers in the same spot while you are driving, but it lurks near the big *C* on cold mornings. When it rises to the big *H*, it is alerting you to stop driving.

HABITAT

One of the prime indicators of the health of the engine and its current operation, the temperature gauge sits directly in front of the driver, usually on the opposite side of the fuel gauge with the speedometer in between.

HOW IT WORKS

The gauge is really showing current in an electric circuit. The current depends on the temperature of the engine, as the circuit includes a component whose resistance to electricity varies with its temperature.

As the car heats up, the resistance of the component decreases, which allows more current to pass through the circuit. As the current increases, it heats a bimetallic strip in the temperature gauge. Like the thermostat in your home heating unit, the bimetallic strip bends when heated. The gauge needle is connected to the strip.

Many cars don't have a temperature gauge. Instead they have a warning light that glows when the engine overheats. A warning light is controlled by an on-off switch that changes based on the engine temperature. You may be able to see this sensor sticking out of the water inlet cover that brings coolant into the engine. It has an electric wire protruding from one end.

INTERESTING FACTS

A high temperature reading can be caused by any of several different conditions. The radiator could be low or out of water. Refilling the radiator should solve the problem. If the low water condition continues to occur, the car needs to be checked by a mechanic for a loose or bad hose, a hole in the radiator, or possibly a faulty water pump. Don't drive when the temperature gauge is pegged, as you will damage the engine.

Tire Pressure Gauge

BEHAVIOR

Forced onto the tire stem, it measures the pressure of air inside.

HABITAT

These are often found in the glove box or center console of the knowledgeable car owner.

HOW IT WORKS

A pin in the end of the tire gauge depresses the release valve in the tire stem letting air flow into the gauge. This occurs only when the gauge is positioned on the stem. The air pressure in the tire forces a piston inside the gauge to move toward the end of the gauge. Connected to the piston is a rod that projects out the end of the gauge. Opposing this movement of the piston is a spring.

More pressure is required to compress the spring farther, which shows a higher pressure on the calibrated rod. Most tire gauges have springs that can be fully compressed when 60 pounds per square inch of pressure are applied.

In 2008 the presidential contest focused for a few days on the efficacy of inflating tires as a way to save gasoline. Although experts disagree, the Department of Energy states that properly inflated tires cut gasoline consumption by more than 3 percent. Since the United States uses about 150 billion gallons of gasoline each year, and with the price approximately $4 per gallon, that tire gauge sitting in your glove box could save us $20 billion annually. You had better get out there and start checking tires!

Toll Transponder

BEHAVIOR

It allows drivers to whiz by the toll booth at almost highway speeds rather than stop to hand over another $2.50 to a smiling attendant.

HABITAT

The transponder often is attached to the inside of the windshield on the passenger side of the car or behind the rearview mirror. Some transponders attach to the dashboard, usually out of the way of the driver.

HOW IT WORKS

Toll transponders are one application of Radio Frequency Identification (RFID) technology. As you drive through a toll lane, an antenna sends out a high frequency radio wave (900 MHz) signal that activates the car-based transponder. The transponder replies with a specific identification code.

Some systems include a metering system in the transponder that you "fill" or top off like a postal meter or Starbucks card—a debit card system. You can go online to top off your transponder by paying with a credit card. Each time you pay a toll, the credit is deducted from the transponder.

Failure to pay or having a low balance transponder causes a bill to be sent to you. Video cameras monitor each lane to get the license plates of any violators.

INTERESTING FACTS

The first electronic toll collection system in the United States was launched in Texas in 1989. Now they are used in many states. Some cities, most notably London and Singapore, discourage cars on over-used center city roads by charging drivers to drive into the city by using these toll-collecting systems.

Turn Indicator

BEHAVIOR

Once pushed up or down, they sing a melodic tune at a beat of about one note per second. They also signal other drivers of your intention to turn. Amazingly, after you complete the turn, they shut themselves off.

HABITAT

Not always visible, they are always within reach on the steering column. They are partially hidden by the steering wheel, but you can find them by reaching under the wheel with your left hand.

HOW IT WORKS

Turn indicators are a marvel of engineering. They are an electro-mechanical device in a largely solid-state world—seemingly anachronistic, but enduring.

Power for turn indicators comes from the battery through the ignition. Thus, when the engine is off, the signals don't work. Emergency flashers, which light up both sides of the turn signals at once, are not powered through the ignition so they can work when the engine is shut off. Power for turn signals goes to a small, cylindrical component called a thermal flasher, which gives the signals their melodic beat.

When you push up or down on the turn indicator a switch connects the circuit so electricity can flow from the thermal flasher to the turn lights on one side of the car. The lights don't come on immediately because inside the flasher one metal spring takes a second to warm up. As current continues to flow inside the flasher, a high-resistance heating element wrapped around the metal spring is heated by electric current. On heating, it expands and bends the spring, which causes a contact switch to close. Electric current now flows to the lights.

Electric current is now diverted away from the heating element and it quickly cools. As it does, it contracts and pulls the spring to open the switch. The turn lights turn off.

But the turn indicator is still in its depressed or elevated position, so current flows to the heating element and the cycle begins anew. Hopefully this cycle doesn't continue too long as you wait at an intersection.

Finally making the turn, you hear a click coming from the steering column and the turn indicator stops. When you first depressed the indicator a pin was extended inside the steering column. As the steering wheel turns it dislodges the pin allowing the circuit to turn off.

UNDER THE CAR

CRAWLING UNDER A CAR isn't always comfortable, but there is a lot to see there. The suspension, brakes, steering, and exhaust systems are visible for your visual exploration. Just be careful of what you touch; it may be hot and certainly is grimy. Oh, and set the parking brake before you climb under.

Brakes

BEHAVIOR

When teaching your children to drive a car, your feet keep reaching for the pedal to activate these. Brakes sap kinetic energy from the moving wheels and convert it into heat. Most would say they slow and stop a car.

HABITAT

Brakes are found either riding above the disc (as in disc brakes) or inside the wheel (drum brakes).

HOW IT WORKS

Disc brakes pinch the disc to slow its motion. Hydraulic fluid is forced from the brake cylinder when you push on the brake pedal. The increased pressure in the brake line forces brake pads on both sides of the disc to squeeze toward each other. The friction of the pads on the discs converts rotary motion into heat. The brake is exposed to air so the heat can dissipate, keeping the brake relatively cool. The pads and cylinders are held above the wheel by the caliper. As the pads wear, they expose a thin piece of metal that rubs against the disc, making that squealing sound that warns you to have them checked.

With drum brakes, often used for the rear wheels, the friction is applied inside the brake drum. Two hydraulic cylinders push the brake lining outward to rub against the drum, which is attached to the wheel. Springs pull the lining away from the drum when you release the brake. Since drum brakes are surrounded by the rest of the wheel, the heat they generate can build up and make the brakes less effective.

INTERESTING FACTS

Both disc and drum brakes were invented in 1902. Louis Renault, the engineering brother of the trio who founded the Renault car company, invented drum brakes, hydraulic shock absorbers, and the turbocharger. Englishman Frederick Lanchester, a giant in car engineering and inventing, invented disc brakes.

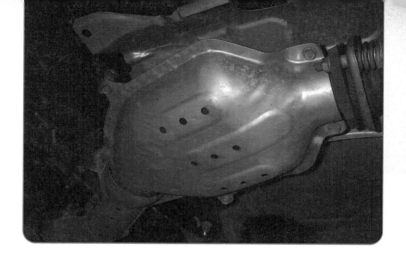

Catalytic Converter

BEHAVIOR

Reduces toxic emissions from cars and trucks by encouraging chemical reactions that break down carbon monoxide (a poisonous gas) into carbon dioxide, convert nitrogen oxides into nitrogen gas (inert) and oxygen, and burn any remaining hydrocarbons in the exhaust.

HABITAT

The catalytic converter interrupts the flow of exhaust gases along the exhaust pipe. It is located upstream of the muffler, beneath the car. It is the bulge in the exhaust system under the middle of the car.

HOW IT WORKS

Catalytic converters work by oxidizing unburned hydrocarbons and reducing nitrogen oxides into nitrogen. Exhaust gases pass through a cylinder filled with a porous ceramic material whose pores give it a huge surface area. The material is coated with platinum, a precious metal catalyst that speeds the chemical reaction between oxygen in the exhaust gases and the hydrocarbons. It strips the oxygen atoms from nitrogen compounds, releasing nitrogen and oxygen, which naturally occur in the atmosphere.

The nitrogen compounds (NO2 and NO) react with oxygen and sunlight to make ozone, which is a major part of smog. Stripping the oxygen from the nitrogen compounds is a process called reduction, the opposite of oxidization that is required to treat the unburned hydrocarbons in exhaust. Thus, the converter has to do two jobs, which are chemically the opposite of each other.

The material used in the catalytic converter has to be extruded (forced through a die with small openings) to make a thin material with large surface area for the exhaust gases to come in contact with. The method of making this has its own patent (#3,790,654). The material used is a ceramic magnesium-aluminum-silicate compound, called cordierite. Corning Glass Works engineers obtained Patent #3,885,977 for cordierite material used in converters.

INTERESTING FACTS

The catalytic process for cleaning up car exhaust was first developed by Eugene Houdry, who earlier had invented catalytic cracking of petroleum. Catalytic cracking greatly increased the quantity of gasoline that could be produced from a barrel of oil. Houdry was inducted into the National Inventors Hall of Fame for this invention. Rod Bagley, Irwin Lachman, and Ron Lewis joined the National Inventors Hall of Fame honored inventors for their pioneering work developing the catalytic converter.

The Clean Air Act of 1970 forced car makers to innovate ways to reduce 90 percent of the emissions from cars. The catalytic converter was developed to meet the stringent requirements of this law.

Coil Spring

BEHAVIOR

Coil springs absorb the bumps that tires encounter and they retard the upward and downward motion of a vehicle as it moves across uneven surfaces. In conjunction with shock absorbers, coil springs form the front suspension of most cars.

HABITAT

Found inboard of the front wheels, springs are mounted around shock absorbers between the upper and lower wishbone or control arms that support the wheel.

HOW IT WORKS

Coil springs are made of special spring steel rods. The rods are heated and wound into the spiral shape.

Coil springs store energy. As the spring is compressed, it stores the energy and then releases it to make a smoother ride.

Springs obey Hooke's Law, one of the basic laws of physics. The law specifies that the position of a body attached to a spring is proportional to the force that pulls or pushes on the body. For example, if you hang a spring from the ceiling, attaching heavier weights to the bottom end will stretch the spring—causing its length to increase. The weight added is directly proportional to the increased length of the spring.

Springs stretched beyond their elastic limit don't obey Hooke's Law. Take a Slinky and pull it far enough and you get a jumbled coil of wire instead of a toy.

Constant Velocity Joint Boot

BEHAVIOR

This rubber housing protects the constant velocity joint hidden inside. The boot keeps out dirt and rocks and keeps in the (molybdenum disulfide) grease (actually a dry lubricant) that lubricates the constant velocity joint.

HABITAT

Found in front-wheel and all-wheel drive cars. Constant velocity (CV) boots are found beneath the car, just inboard of the front wheels.

HOW IT WORKS

The rubber boot provides a protective environment for the CV joints hidden inside. The joint itself is more interesting.

CV joints allow the front wheels to receive power from the axle while the wheels turn side to side and ride up and down over a rough road. The joint transfers rotational energy from the drive shaft to the wheel.

Since each wheel moves up and down but the drive shaft connected to the transmission needs to stay stationary, there has to be a flexible joint between the two. A universal joint would work, but it has the problem of varying the speed as the shaft turns. A CV joint solves this problem. It provides a consistent speed of rotation regardless of the angle between the two shafts. The CV joint is needed for the front wheels of a front-wheel drive car because the angles between the two shafts can vary so much (30 degrees), which would make the output rotational speed too erratic.

The CV joint is a ball-and-socket joint where one shaft is the ball and the other ends in a socket. Six steel balls ride in a race (grooves cut into the ball-and-socket) and as the two axles turn, the balls move back and forth. The flexibility afforded by the sliding balls allows the joint to turn with constant speed.

INTERESTING FACTS

An engineer at Ford in 1926 invented the constant velocity joint. Alfred Hans Rzeppa was awarded Patent #2,010,899 in 1935.

Differential

BEHAVIOR

The differential allows the nonsteering wheels to rotate at different speeds so the car can corner without putting undue wear on the tires. The wheel on the inside of a turn moves a shorter distance than does the outer wheel. If the axle doesn't allow the wheels to turn independently of each other, the tire of one wheel will be dragged across the ground.

HABITAT

You can easily see the differential beneath the rear end of a car. It is the large metal bump in the middle of the axle.

HOW IT WORKS

The axle isn't one continuous metal bar but instead is composed of two half-shafts, each connected to a wheel on the outside and a sun gear inside the metal bump. Between the two sun gears is a system of small pinion gears. A sun gear is one that is located in the center of a series of other gears, called planetary gears, that revolve around it.

The drive shaft terminates in another pinion gear. This gear turns a large, vertically orientated crown wheel inside of which are the sun and planetary gears.

When you drive in a straight line, the engine spins the drive shaft, which turns the crown wheel. The two pinions do not spin and each of the sun gears receives the same torque or turning power. When you make a turn, the pinion gears rotate around the axles allowing the two wheels to turn at different rates, while still supplying each wheel with torque.

INTERESTING FACTS
Differential gears were used in ancient calculators, carts, and watches long before cars were invented.

Gas Tank

Provides a reservoir for the fuel that keeps your car on the go.

It is usually found beneath the car, close to the filling tube. However, manufacturers have placed it in many different locations.

Gasoline flows into the tank through a filler neck by gravity. Tanks hold at least 8 gallons and rarely more than 25.

The tank is made of thin steel but has raised ridges to provide additional strength. The steel is coated with a lead-tin coating to protect it from rust, and in some models it also has an undercoating for additional protection. On some cars the tanks are made of either aluminum or polyethylene instead of steel.

Inside the tank the gas is prevented from sloshing around by baffles. The baffles are sheets of steel (or whatever the tank is made of) that span the width of the tank. They have holes along the bottom so the gasoline can flow slowly into the different compartments but not slosh back and forth.

Also inside the tank is a float sensor that connects to the fuel gauge on the dashboard. As the fuel level rises and falls, the float moves with it. It is connected to a variable resistor that controls the current in the electric circuit that controls the fuel gauge.

Some cars have submersible fuel transfer pumps that push the fuel up toward the engine. These are electric pumps. To prevent foreign matter from being pumped into the engine, the intake side of the fuel transfer pump has a straining screen. The pickup tube or hose is positioned about a half-inch above the bottom of the tank so it doesn't draw in any debris that has settled out.

Tanks have to breathe, or let air in, as they are filled and emptied. To prevent gas fumes from escaping into the atmosphere, a filter is inserted into the tube that admits air.

Jack

BEHAVIOR
Lifts the weight of the car off one wheel so a tire can be changed. On a dark and stormy night, this is a truly uplifting experience.

HABITAT
It spends nearly all of its life resting comfortably in the trunk, most often in a special compartment with other tools and possibly the spare tire. When pressed into service it latches onto the frame of the car just forward of the rear wheels or just behind the front wheels.

HOW IT WORKS
Portable jacks found in cars are generally of one of two types.

A screw-turned scissors lift raises the car as the operator turns a long handle. The handle rotates a screw that pulls opposite sides of the scissors lift together.

A ratchet jack requires the operator to press down with considerable force on the handle of the jack. This leverages the car up a fraction of an inch and sets a pawl or spring-loaded finger to engage the ratchet to prevent the car from sliding down. To let the car down, the operator pushes a lever that reverses the direction of the ratchet. Now each push on the handle lowers the car to the next lower notch.

Jacks at garages are generally hydraulic. The operator uses a long lever to increase the pressure inside a small piston. The piston is connected to a larger piston that lifts the car. Since the pressure is the same throughout the closed system, the small downward force on a small piston can lift a much larger force or weight resting on a larger piston. Pressure is force divided by the area, so a small force in a small area has the same pressure as a larger force in a larger area. To prevent the car from pushing the larger piston back down, a valve in the system only lets hydraulic fluid flow in one direction. To let the car down, this valve is opened.

Leaf Springs

BEHAVIOR
They bound up and down with each bounce in the road, absorbing some shock and providing a smoother ride.

HABITAT
Leaf springs are most often found supporting the rear wheels of cars and trucks. The axle is attached near the center of the leaf spring and the car frame is attached to each end.

HOW IT WORKS
Leaf springs are several steel bars of different lengths that are joined together and held in place by metal bands. The spring is curved, and when weight is applied to the spring it stretches and straightens the spring. At the center of a leaf spring it attaches to the axle with a long U-bolt. The ends of the spring are bolted to the bottom of the car's frame.

When a leaf spring is compressed, the leaves slide past one another, potentially making an annoying noise. To prevent the noise, nylon or rubber pads are place between the ends of the leaves. Rebound clips are located along the leaf spring to help hold the leaves together as it rebounds from compression.

INTERESTING FACTS
Leaf springs were invented in ancient Egypt and used for launching projectiles at enemies. Before cars were introduced, leaf springs were used in wagons and coaches.

Muffler

BEHAVIOR
Cuts the noise coming from car and truck engines. Unless, of course, some young driver cuts holes in it.

HABITAT
The muffler lives at the end of the combustion process, beneath the back end of the car.

HOW IT WORKS
Exhaust gases are piped into the muffler at elevated pressures and then pass out the tailpipe. The escape of high pressure exhausting into the atmosphere generates noise.

Noise is suppressed inside the muffler by destructive interference. Sound energy reflects off the inside of the muffler and tends to cancel itself out. This is the principle of noise-canceling headphones: two identical sound waves that are exactly out of phase with each other can cancel each other.

Exhaust gases enter the muffler's first chamber where they escape into a resonating chamber. The resonator is designed to reflect sound waves back at incoming waves so the two waves interfere with each other.

Exhaust and sound escape to a second chamber for further dampening. The exhaust has to flow through small holes in the pipes inside the muffler, thus reducing the noise further.

The body of the muffler is a sandwich of metal with a layer of insulating material between them. This helps noise reduction by absorbing sound.

A different design for a muffler is the straight-through or glass pack muffler. Combustion gases enter through a perforated pipe inside the muffler. The gases escape through the perforations into a chamber filled with sound-absorbing material (fiberglass, for example) and then out the tailpipe. The straight-through design reduces back pressure, making the engine more efficient.

The design of mufflers is a balance between suppressing noise and reducing power output by the engine. Like sticking a banana up the tailpipe, a muffler creates back pressure, which retards exhaust gases and decreases engine output. Too much exhaust restriction in the muffler will cause noticeable reduction in power.

INTERESTING FACTS

The silencer or muffler was invented by Hiram Percy Maxim. He used the same technology to invent the silencer for firearms. His father, Sir Hiram Stevens Maxim, was a more famous inventor. He invented the machine gun and the modern spring mouse trap, and had pre-Wright Brothers success getting a steam-powered airplane to take off.

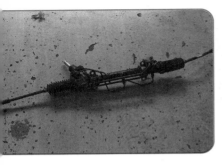

Rack and Pinion Steering

BEHAVIOR

It connects the steering shaft, which is controlled by the driver turning the steering wheel, with the turning wheels. It converts the rotary motion of the steering wheel into the side-to-side motion needed to turn the wheels.

HABITAT

It resides at the bottom end of the steering column connected to the steering shaft.

HOW IT WORKS

As you turn the steering wheel, it rotates a round gear called a pinion. As the pinion rotates it pushes a flat gear called a rack. Acting together, the rack and pinion converts the rotational motion of the steering wheel into motion to the left and right. The ends of the rack move the tie rods that link the motion to the steering knuckles that hold the wheels.

The gearing of the rack and pinion increase the driver's force in turning, although not as much as some other steering mechanisms. However, rack and pinion steering provides a better "feel for the road," as the motion of the steering wheels is transmitted to the driver.

In this photo, the rack and pinion is encased in a metal cover. On each end are the two tie rods. The metal bar protrudes upward from the pinion gear and connects to the steering shaft.

INTERESTING FACTS

Rack and pinion steering is relatively new. An Australian, Arthur E. Bishop, invented the rack and pinion variable steering gear in 1973 and obtained U.S. Patent #3,753,378.

Resonator

BEHAVIOR
It reduces the engine noises coming out of the exhaust system.

HABITAT
It is a component of the exhaust system found on more expensive cars. It is connected to the exhaust pipe just upstream of the muffler and downstream of the catalytic converter.

HOW IT WORKS
Think of the resonator as a small muffler. It reduces engine noises much the same way that a muffler does.

Like a muffler, a resonator can either pass the gases straight through or cause them to reverse the direction of their flow (reverse-flow design). As the gases change direction in a reverse-flow design, back pressure builds up reduces the engine's efficiency. So sound suppression must be balanced against the loss of engine efficiency.

A resonator is a largely empty enclosure that cancels sound waves of a particular frequency. In physics circles it would be called a Helmholtz Resonator.

As sound enters the resonator along with exhaust gases, some of the sound reflects off the interior walls and bounces back toward the next set of sound waves. There the two waves interfere with each other—destructive interference—reducing the sound level.

Resonators cannot reduce sounds of all frequencies and so are designed to reduce sounds at frequencies generated when the engine is making the most noise.

Roll Bar (a.k.a. Anti-Roll Bar or Sway Bar)

BEHAVIOR
It reduces the lean a car will make in a turn, and it improves the steering characteristics.

HABITAT
The roll bar is a long, generally U-shaped rod of steel that connects the wheel on the right side to the wheel on the left side.

HOW IT WORKS
The steel rod of a roll bar acts like a spring. However, rather than a coil or leaf spring, it is a torsion spring; it twists under pressure. Its resistance to twisting provides the stability from side to side.

As a car enters a turn, weight is shifted toward the outside wheel. The car leans outward, dropping down on the outside and rising on the inside. Going too fast and making too strong a turn can cause a car to roll over to the outside.

The roll bar resists the tendency of the car body to lean, providing a smoother ride. As the car leans to one side, the roll bar is twisted. The steel resists this twisting and tries to return to its original untwisted position.

A disadvantage of having roll bars is that the road bumps felt by one wheel are carried to the other wheel by the bar. This can make the ride even rougher on a bad road. Some cars have computer controlled systems to overcome this problem by hydraulically adjusting spring height.

Shock Absorber

BEHAVIOR

They smooth the ride. Without shocks (or struts, which substitute for shocks), each pothole would launch the car vertically, giving rise to Slinky-like up-and-down gyrations. Shocks dampen the vertical bouncing as the car hits holes. Undamped spring motion makes driving much more dangerous and downright uncomfortable.

HABITAT

Shocks separate the wheel axle and the frame of the car. Shocks are often surrounded by, or are inside of, coil springs.

HOW IT WORKS

Shock absorbers reduce the vertical oscillation of springs. Holding one end of a Slinky and releasing the other end will set up a long-lasting up-and-down oscillation—fun to watch but annoying and dangerous if your car does it. Shock absorbers take some of the spring's energy and dissipate it so the spring doesn't rebound as energetically.

Shock absorbers are sealed cylinders filled with oil with a piston inside. As the wheel bounces up the shock absorber (and spring) is compressed, driving the piston into its cylinder. The piston displaces oil that is squeezed through openings that slow the piston's movement, thus absorbing the shock. As the wheel moves down, the absorber lengthens and the piston withdraws farther from the cylinder. Now oil flows back into the cylinder and its movement slows the extension of the shock absorber.

INTERESTING FACTS

The front door of your home probably has a shock absorber. Most outer storm doors have piston devices that slow the doors' closing so they don't slam shut.

Springs

Springs are part of the suspension system that holds the chassis to the wheels. They help cushion the ride by resisting the vertical motion of the car.

HABITAT
Springs reside beneath the car toward the inside of each wheel.

HOW IT WORKS
Springs are made of hardened spring steel so they can bend and return to their original shape. Their job is to compress under load and rebound.

Three kinds of springs are often seen beneath cars and vans on the road today. Leaf springs (shown here, behind the shock absorber), once popular on all four wheels of cars, are now used mostly on the rear end of cars and on heavier vehicles, as they spread the weight load over a larger area of the chassis. Vehicle leaf springs were invented in the 16th century to cushion the ride of carriages. They consist of several flat bars of steel held together with clamps. The bars vary in size, the smallest ones being farthest from the axle. The axle is attached to the center of the leaf spring, which is attached at each end to the car frame.

Coil springs are made of steel wire wrapped into a helical shape. The coil springs in cars resist being compressed between bouncing wheels and the chassis. Coil springs can be used independently or in combination with shocks—the combination is called a strut.

Struts

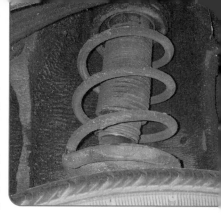

BEHAVIOR
They make your ride smoother while holding the chassis above the axle.

HABITAT
They inhabitant the dirty world under your car, just to the inside of the wheel hub. Look for the coil spring with a piston-like device inside.

HOW IT WORKS
Struts dampen vertical motion of the car. Push down on the front of your car, and when you release your weight the car should rebound and stop. Without a dampening system, the car would continue to move up and down as the spring lengthens and shortens, like a Slinky. Driving down the road with bad struts or shock absorbers is a bad and dangerous ride.

A strut is a combination of a coil spring and shock absorber. The spring wraps around the shock absorber. Acting together they reduce the vertical motion of the car while holding the wheels to the chassis.

INTERESTING FACTS
MacPherson struts are a popular suspension system on the front end of cars. Earl MacPherson is credited with designing the struts first used on production cars in the late 1940s. But it isn't the only strut available; it's just the best known.

Tailpipe

BEHAVIOR

It channels exhaust gases from the exhaust system into the atmosphere. It's the demarcation line for gases from being engine exhaust to becoming air pollution.

HABITAT

Tailpipes on cars are beneath the rear bumper. On large trucks, the end of the exhaust system can be located by the cab, pointing skyward.

HOW IT WORKS

Tailpipes are steel tubes that direct the exhaust gases out away from the car. Some are welded to the muffler and some may have a built-in resonator.

INTERESTING FACTS

That white plume that you see emanating from the tailpipe of cars and trucks isn't pollutants. It's water vapor. When engines first start and are cold, water vapor from the engine's exhaust cools and condenses in the tailpipe and leaves as a cloud of fog. After a few minutes the tailpipe heats up enough so the vapor doesn't condense and the plume disappears.

Years ago when I was conducting research in the Antarctic we had to keep our vehicles running to prevent them from freezing up and leaving us stranded miles from base. One day our team leader burst into the tool shed, grabbed an ice auger, and ran outside. I followed to see him jam the auger bit into the tailpipe of the truck and start turning to auger out the accumulated ice. Anything that blocks the exhaust gases—ice, snow, bananas—can stall the engine and endanger the passengers. Carbon monoxide gas can escape from a blocked exhaust system and enter the passenger compartment. If your car has been in blowing snow, check the tailpipe.

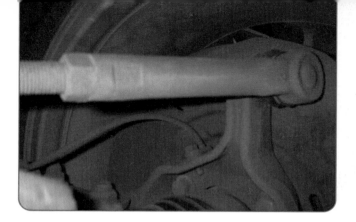

Tie Rod

BEHAVIOR
Connects each front wheel to the steering mechanism (the rack in a rack and pinion steering system) so the car can turn.

HABITAT
They connect to the steering knuckle of each front wheel. The knuckle holds the wheel and attaches to the tie rod while letting the wheel rotate in a turn. You can identify the tie rod as a steel rod that connects to both wheels. The outer ends are threaded so they can be bolted onto the knuckles.

HOW IT WORKS
The two front wheels have to be both connected to the steering system and have to be parallel to each other. If they are not parallel, one tire or both will drag on the ground producing uneven wear and shortening its life. This condition requires a front-end alignment.

The tie rod is what connects the two wheels to the steering rack. It connects to either the rack gear or link (depending on the type of steering system the car has). The outer ends connect to the wheel through a device called a knuckle (similar to a hinge) that allows the wheels to rotate inward and outward while attached to the tie rod.

Adjusting the tie rod is one of the three adjustments made to align the wheels. The effective length of the tie rod adjusts the "toe" of the front wheels. If it's short, the wheels "toe in," which means that they are not parallel to each other with the front tires being closer together than the back. "Toed out" is the opposite situation. Both promote excessive wear on the tires.

INTERESTING FACTS

There are three adjustments in a wheel alignment. One, camber, deals with the tires in the vertical plane. Viewed from the front of the car, do the top of the tires lean in or out? If they do, the tires have camber. Caster refers to the angle of the axis that the wheels pivot on when turning. Is this line vertical or orientated forward or backward? In designing a car the caster angle determines a balance between the effort required to steer, the stability at high speed, and how effective the car is in turning. The third adjustment affects the toe in or out and it is made to the length of the tie rods.

Tires

Tires support your car, help hold
the car to the road, and smooth out
the small bumps in the road. You
don't appreciate what they do until
they leak air and quit doing it.

This is where the rubber meets the road! They are the rubber meeting
the road. Attached to each wheel, tires hug the road and maintain contact between car and highway. Tires wrap around each of the four
wheels that support the car. One more, a spare, should be mounted on
a wheel and stored in the trunk or beneath it.

The pressure for your tires is probably around 40 psi, or pounds per
square inch. It doesn't seem possible that 40 pounds of pressure can
support a two ton car, but it does. Pressure is force per area so you
multiply the tire pressure by the area of contact between tires and road
to get the total weight that each tire supports.

I measured the footprint of one of my tires and it was about 8 inches
wide by 10 inches long. (It was not, I confess, a very accurate measurement.) For all four wheels, that's 320 square inches of connection
between the road and car. At 40 psi, the tires could easily support six
tons, more than twice the weight of my car.

Tire parts include the tread, the sidewall, and the bead. The tread is
a high friction layer of rubber that lies on the outer circumference of
the tire. It has a pattern of grooves cut in it to allow water on the road
to escape to avoiding hydroplaning and to grip the road. The bead is
the inner edge of the tire. It makes contact with the rim and provides
the seal that maintains the tire pressure. The sidewall lies between the
other two parts. It consists of several layers of material protected by
an outer covering of rubber. The body of the tire is made of crisscross-
ing belts made of steel, fiberglass, or synthetics. The air pressure

inside the tire exerts tension on the tire materials that actually support the weight of the car. The rubber sidewalls and tread lie on top of belts of fabric, initially rayon and more recently nylon or polyester. Run-flat tires have thicker and heavier sidewalls to support the car even when the air has escaped.

Radial tires have belts of fabric cord aligned with a radius of the tire. In this alignment the cords directly oppose the outward forces of spin. Bias-ply tires have the cords aligned at a diagonal. Radials also have radially aligned belts of steel or fabric between the cord fabric and the rubber tread. Even with low air pressure radial tires don't sag as much (as bias-ply tires) until they are very low on air, so you can't rely on visual inspections to know if they need refilling. Get them checked often.

Tubeless tires are held in place on the wheel by bead assemblies at the inner edges of the tires. This lump of material at the inside edge runs around the tire. Air pressure forces the bead assembly outward, sealing the edge of the tire against the wheel.

INTERESTING FACTS

Rubber tires became possible after Charles Goodyear's discovery of vulcanization in 1839. Inflated rubber tires were the invention of John Dunlop in 1888, whose primary concern was bicycle tires. Tubeless tires were introduced by B. F. Goodrich after their patent in 1952.

In the 1950s buying tires for your car was an annual event: bias-ply tires lasted only about 15,000 miles. Fast driving on the new interstate highway system caused even faster tire wear due to uneven wear at higher speeds and the abrasive nature of the sand used in highway concrete.

Radial tires are a huge improvement over the older bias-ply tires. The term radial was introduced by tire maker Pirelli. Like many technologies, radials were invented long before they were adopted. The first patent was issued in 1914. But foreign-made radials weren't introduced to the United States until the 1960s. American bias-ply tire makers had changed the way they made tires to save money. They had reduced the thickness of the sidewalls by half, and that greatly increased tire failure and the public's displeasure. In response, Sears began offering Michelin-made radial tires in 1966. Michelin had developed steel-belted radial tires in the late 1940s and early 1950s, but were slow to export them to the United States. This innovation doubled tread life, cut fuel consumption, and made driving safer. By 1975 nearly 90 percent of the new cars sold in America had radial tires.

Those numbers along the side of the tire—such as "P215/65R15"—do you know what they mean? First, the leading P designates the tire is used on passenger vehicles. The tire in your trunk might have a T for temporary, and if you drive a truck or sports vehicle the tires might say LT for light truck.

The first number specifies the width of the tire in millimeters. So 215 is 215 mm wide, or about 8 inches. Next is the height of the tire from the outer tread to the inner circumference. But just to confuse you, this measurement is given as a percentage of the width. So 65 shows that this tire has a height 65 percent of the tire width.

The letter that follows, R, shows that the tire is a radial tire. That is by far the most common type used today. Following the tire type is the rim width, measured in inches. In this case, 15 inches.

Following the rim width you might find a series of numbers and letters specifying the quality of the tire. These may or may not appear on the tire, but may be on the receipt when you purchase new tires. First is tread wear, specified as a number. Then comes traction with AA being the best and C being the least best. Then comes a temperature dissipation rate of A, B, or C. An A rating means that the tire is effective in preventing heat build-up that can damage a tire. The last are load and speed ratings, for which you need tables to interpret. These numbers don't show up on my tires.

Transfer Case

BEHAVIOR

It is a gearbox that distributes power from the transmission to wheels on both front and back axles in four-wheel drive cars.

HABITAT

The transfer case can be found beneath cars that have four-wheel drive. It is directly behind the transmission and may be built into the transmission. The other option is for it to have a short driveshaft separating it from the transmission.

HOW IT WORKS

With only one engine and four wheels needing power there has to be a device that directs power to each wheel. The transfer case is it.

The distribution of power within the transfer case is done either by gears or chain drives. In cars with part-time four-wheel drive, the driver selects two or four-wheel drive with a shift lever, similar to a manual transmission lever that connects to the transfer case. The transfer case can also allow drivers to select high torque/low speed option for serious off-roading or low torque/high speed for normal driving.

Most four-wheel drive cars have chain drive for the front wheels.

Today we don't associate four-wheel drive vehicles with sports cars. But the first internal combustion vehicle with four-wheel drive was a sports car. Dutch brothers Jacobus and Hendrik-Jan Spijker built the six cylinder, 60 horsepower *Spyker* as a racing car in 1903.

Universal Joint (U-Joint)

BEHAVIOR
A universal (or U) joint transfers rotary motion between two shafts that are not in line with one another.

HABITAT
U-joints are found beneath cars, connecting the driveshaft to the transmission and differential.

HOW IT WORKS
Looking at a U-joint you can see that the two shafts it connects end in a U-shaped yoke. The two yokes fit together 90 degrees apart. Holding them together is a cross-shaped piece of metal called a spider. Each end of the spider's arms fit into a hole in one of the sides of a yoke.

With increasing angle between the two shafts, there is an increasing variation in the speed. The speed changes twice per revolution of the shaft. Think of the case where the two shafts are nearly perpendicular to each other. The output shaft would have a jerky motion. Constant velocity joints are a type of universal joint that eliminates this problem of changing speed.

INTERESTING FACTS
The idea for universal joints grew out of gimbals (pivoted supports), which had been used for thousands of years. The first use for transmitting power was demonstrated by the scientist Robert Hooke in 1676. Hooke is known to physics students for his law on elasticity, Hooke's Law. Henry Ford gave the universal joint its name.

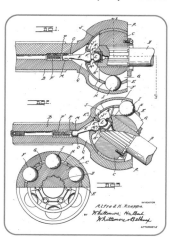

Wheel

Wheels hold the tires onto the car.

Find a tire on a car and you'll see a wheel in the center.

Wheels are made of stamped steel. They are made in two parts. The outer rim is welded to the inner or center section. The inner section has four to six holes to mount the wheel to the hub.

The rim is bolted onto the rotor for disc brakes or to the brake drum for drum brakes. Either lug bolts are threaded through the holes in the wheel into the hub or the hub has wheel lugs that project outward through the holes in the wheel. Lug nuts screw onto the wheel lugs. These nuts or bolts are usually covered with a hubcap.

Contrary to popular opinion, the wheel was not human's first invention. People were using spears, bags, clubs, and all manner of other devices centuries before the wheel was invented. And they were used first in making pottery—wheels for carts weren't used until 5,700 years before the present day in ancient Mesopotamia.

Wheel Clamp (or Denver Boot)

BEHAVIOR

The boot is applied by parking authorities to the front wheel of any unfortunate soul who is caught not paying his or her parking fines. It prevents a car from being moved. Some people use them to ensure that their vehicles or trailers are not stolen.

HABITAT

Hopefully, it never is found on one of your wheels. You see it more frequently on cars in major cities.

HOW IT WORKS

It fits around a tire and wheel and is locked in place with a padlock. Driving a car with a boot installed will damage the car and make the car uncontrollable. The clamp covers the lug nuts of a wheel so the car owner cannot replace the wheel and drive away.

Boots weigh about 20 pounds and can be applied in less than a minute. Taking one off without the key, however, takes much longer.

INTERESTING FACTS

A violinist invented the Denver boot. Frank Marugg, a violinist with the Denver Symphony Orchestra, invented it in 1953. Having friends in the Denver city government, he got the city to use the boot to improve enforcement of parking laws.

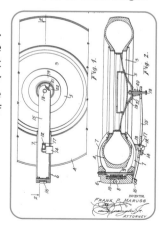

5 UNDER THE HOOD

WHAT NOISY BEAST IS IT that resides beneath the hood of your car? It breathes air, consumes petroleum, and belches particulate-laden exhaust responsible for all manner of undesirable environmental and health effects.

For all its negative attributes, few machines have gained such widespread acceptance. Anywhere you go in the world people are using gasoline engines to move themselves and their goods, to move water, and to make electricity.

INTERNAL COMBUSTION ENGINES

Gasoline engines can convert about 25 to 30 percent of the energy in burning fuel to moving the vehicle. Diesel engines operate a bit more efficiently at up to 40 percent. The remaining or wasted energy is lost as heat. Yet even at these low ratings, internal combustion engines are the right choice for many applications.

Internal combustion means that the explosions that power the engine occur inside the engine, in cylinders. In steam engines, the combustion occurs outside the engine in a separate fire box.

Air and gasoline are squirted into the cylinder in a ratio of about 15:1. That is, 15 parts of air mix with one part of gasoline. This mixture is compressed by a piston moving upward in the cylinder. At just the right moment in the cycle, a spark plug fires and ignites the mixture.

The explosion drives the piston downward, and as it moves it rotates the crankshaft. One cylinder operating a crankshaft makes for a rough-operating engine so usually cars have at least four cylinders. The additional cylinders not only smooth out the motion, they also provide more power. Eight are even more powerful, but use prodigious quantities of fuel.

Valves above the cylinder (overhead valves) let air and fuel into the cylinder as the piston moves downward. They close before the piston begins its upward stroke. Other valves open to let out the exhaust gases resulting from the combustion. These valves may open and close 50 times each second. Strong springs return the valves after being lifted by the cams.

This describes how most gasoline engines work. Most use this Otto cycle, named for its inventor, Nikolaus Otto. A more recent variation of the Otto cycle was invented by Ralph Miller and is called the Miller cycle.

Miller cycle engines have superchargers that force air into the cylinder. Rather than close the intake valve while the piston is compressing the air/fuel mixture, the valve is held open for about 20 percent of the compression cycle. During this period, the piston doesn't have to use as much energy to compress the fuel/air mixture in the cylinder, so each cylinder generates nearly the same energy but expends less energy getting it. Further, the supercharged air is cooled (by a device called an intercooler). The cooler air allows the timing of the spark to be delayed and the resulting compression to be higher. These changes provide another boost in engine efficiency. Mazda uses Miller cycle engines in some of its cars.

A diesel engine works pretty much the same way as an Otto cycle gasoline engine, except that it uses a heavier fuel and doesn't use spark plugs. Instead of a spark causing the explosion, the high pressure of the piston compressing the fuel-air mixture causes ignition. Air enters the diesel engine from a valve and is compressed. In a diesel engine the air is compressed up to twice as much as in a gasoline engine. When the piston is at the top of its stroke and the air inside the cylinder is about 1,500° F, the fuel is sprayed into the cylinder.

Bang! The piston is driven downward powering the crankshaft. Although diesel engines don't have spark plugs, some have glow plugs to warm the cylinders on a cold start.

Of course, engineers could not let gasoline and diesel engines go without tinkering with them. Their automotive creativity manifests itself in a variety of engine types. The Hemi engines lauded by Chrysler Motors has a hemispherical or domed combustion chamber rather than a flat head over the chamber. The shape improves the mixing of fuel with air to get more kick from each explosion. Million dollar ad budgets aside, most gasoline engines today have hemi-like combustion chambers that differ little from the vaulted Hemi.

The rotary or Wankel engine has rotary pistons that spin around in a circle. Rather than the violent vibrations of the reciprocating piston motion in other engines (up, stop, down, stop), the rotary pistons spin smoothly with no stops throughout the combustion cycle. The rotor spins around a shaft and gives it power. Each revolution of a rotor delivers one set of combustion explosions and one pulse of power, rather than one pulse for every two strokes of a traditional (four-cycle) combustion engine. To ensure complete combustion rotary engines typically have two spark plugs for each rotor. Mazda has offered several models.

ELECTRIC MOTORS

Before internal combustion engines were popular in vehicles, people were driving electric cars. Now, a century later, we are looking again at the advantages of electric cars.

Unlike most cars that burn gasoline or diesel fuel to generate heat and motion, electric cars use energy stored in batteries to power motors. The chemical reaction of batteries is reversible so batteries can be charged and discharged many times. One benefit of this system is the reduction of exhaust gases in crowded cities. Instead, any pollutants are released at the site of the electric generator, where hopefully they can be controlled more effectively. Electric cars are less expensive to operate, but their initial cost, largely the cost of the batteries, discourages many buyers.

Manufacturers are using a variety of battery types in electrics. Some use the lead-acid batteries that gasoline engine cars use, but electrics require many more of them. These are very heavy but inexpensive—at least in relation to the alternatives. More practical are nickel metal hydride, but they cost much more. They can increase the car's range, and they might last as long as the car does, but their high cost is prohibitive to many.

In electric cars the accelerator pedal is connected to an electronic control system that interprets the position of the pedal and increases or decreases the voltage carried to the motor. The motor can be either AC or DC. An AC system requires the conversion of the DC power from the batteries into AC current to run the motor. DC motors are often the same ones used in forklifts.

Electric cars can recapture some of the car's kinetic energy to generate electricity. When the car is slowing down, the car's momentum keeps it moving and the motor turning. The motor then acts as a generator, able to recharge the battery.

HYBRID MOTORS

Hybrid cars use electric motors but also have gasoline engines to recharge the batteries when needed. There are several types of hybrids. Toyota's Hybrid Synergy Drive uses two motor-generators and a gasoline engine. A motor-generator can operate either as a motor, when electric power is supplied to it, or as a generator of electricity, when mechanical power is applied.

One of the two motor-generators is mounted on the front transaxle. At slow speeds, nickel-metal hydride batteries provide power to the motor generator on the drive shaft. At higher speeds, about 40 mph, the gasoline engine kicks in to add power to the wheels. The engine also turns the other motor-generator to generate electricity that can either recharge the batteries or provide power to the motor-generator on the axle for additional power.

In this design, there is no starter for the gasoline engine. The motor-generator that is turned directly by the engine acts as the starter. Initial power is provided by the batteries.

As the car accelerates, the gasoline engine and the axle-mounted motor generator provide the power. When the engine is producing more power than needed to drive the car, it generates electrical power through the second motor-generator. When the engine needs help getting the car up a steep hill, the motor-generator on the axle can assist. It draws power either from the battery or from its partner motor-generator. Going down a steep hill the car can capture some of the potential energy through the motor-generator mounted on the axle.

To go in reverse, rather than shift gears the axle-mounted motor-generator receives electric power with the opposite polarity, so the motor runs in reverse. The gasoline engine isn't used in backing up.

Selecting the right combination of battery, motor-generators, and engine is the job of a computer. Drivers don't control the engine directly, they make inputs into the computer that controls the motor-generators and engine. If the computer quits, so does the car. But the advantage is greatly increased fuel efficiency and quieter operation.

From a car engine standpoint, these are exciting times. A wide variety of engine technologies are vying for marketplace approval and it's impossible to say with certainty which will dominate. But judging from the past we know that from the many competing technologies only one or two will prevail and the rest will be relegated to the history books.

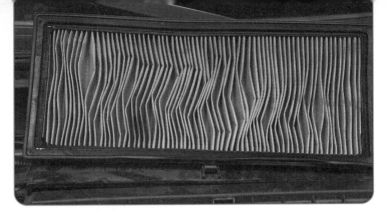

Air Filter

BEHAVIOR
Air filters remove much of the particulate load in the air, keeping it out of the engine. Dirt in the air could clog small openings in the engine, restricting the flow of air or abrading parts.

HABITAT
Air filters sit directly above the engine.

HOW IT WORKS
Most air filters today are made of resin-impregnated paper supported by a rim of plastic with a urethane gasket. The paper is folded or pleated to create a large surface area. Some filter elements have dimples to further increase the surface area so more particles are arrested. Incoming air has to travel through many sheets of paper before entering the engine's intake valves.

Older cars used oil bath filters. In these filters, larger particles are thrown into the oil bath where they are trapped. The oil bath needs to be changed periodically. Smaller particles are caught in a fibrous material that surrounds the oil bath.

INTERESTING FACT
For every gallon of gasoline consumed about 10,000 gallons of air are sucked through an air filter.

Alternator

BEHAVIOR

It converts mechanical energy that the engine produces into alternating current to run the car's electrical system. Older cars had generators that produced direct current and filled the same role.

HABITAT

It is found on the side of the engine. A rubber belt from the crankshaft pulley turns the alternator.

HOW IT WORKS

Alternators make alternating current (AC) by spinning a magnetic field. Coils of conducting wire surround the spinning magnets. Electrical current is inducted in the coils as the magnets spin. The direction of the current changes during every rotation of the magnets to produce AC.

Alternators replaced generators because they can be made stronger, lighter, and less expensive. They are easier to turn than generators and have a smaller pulley so they spin two to three times faster than the engine itself. (The pulleys and belt operate as a gear system that speeds up the rotation of the alternator.)

To charge the battery, current from the alternator is converted into direct current. A diode rectifier does this by limiting the direction the current can flow. The advent of solid-state diodes in the 1960s allowed the transition from generators to alternators. A voltage regulator controls the generator of electric power.

When you turn the ignition key the battery light comes on. The light is part of a circuit that sends a current to the alternator windings to start the magnetic field. As you turn the engine on it spins the alternator, which now generates electricity. But if the light remains on, it is indicating that the alternator isn't producing enough electric power. It could be that the belt that turns the alternator has broken or that the alternator itself is failing. Of course the car will still run, but you might notice that your headlights get progressively dimmer. Once you stop the car you probably won't be able to restart it, as the battery will be dead.

Under normal operations the light on the dashboard goes out after a few seconds. As the alternator starts generating electric power, it sends an opposing current to the light in the dash causing it to go out.

INTERESTING FACTS

The switch from using generators in cars to using alternators came about in the 1960s when solid-state diodes could be produced inexpensively. Diodes are needed to rectify the alternating current into direct current to charge the battery. With cheap diodes available, car manufacturers switched to the less expensive and more durable alternators.

Battery

BEHAVIOR

They provide the cup of Joe in the morning to start the car. Batteries store chemical energy and convert it into electric energy to power the starter and the many electrical appliances in a car.

HABITAT

In most American-made cars, batteries reside under the hood at a high level so they are accessible.

HOW IT WORKS

Most car batteries are lead-acid, wet cell batteries. The cells are contained inside a polypropylene case.

The battery has six cells inside, each of which generates about two volts of electricity. The six cells are connected in series to yield 12+ volts. In a series circuit, the negative terminal of one cell is connected to the positive terminal of the adjacent cell and the voltages of each cell are added together to give the total voltage of the battery.

The cells have plates that serve as electrodes. The plates are made of lead and lead oxide and they are immersed in a bath of diluted (36 percent) sulfuric acid.

When the battery is discharging a chemical reaction occurs that converts lead and lead oxide into lead sulfate and releases electrons that comprise electrical current. When the battery is charging (from the alternator) this chemical reaction is reversed so the lead sulfate becomes lead (at the negative electrode or plate) and lead oxide (at the positive electrode).

The positive terminal of the battery is connected to the starter motor. The negative terminal is connected to the car frame with a large wire. The ignition switch completes the circuit and powers the starter.

Jump-starting the car is in essence connecting in parallel the dead battery to a battery in an operating car. Thus, the working car's battery powers the other car's starter. It's important to know that the negative side of the battery connects to the frame. Touching the positive connection of the live battery to the frame will cause sparks to fly.

Most electric cars use the same lead-acid batteries that gasoline-powered cars use. However, they use many batteries instead of one. The lighter-weight alternative to a lead-acid battery is a nickel metal hydride battery.

INTERESTING FACTS

Before "maintenance-free" batteries, checking the water level in your battery was part of the car maintenance ritual. Now you never add fluid to the battery. Improved design for batteries reduces the loss of water. And solid-state electronic controls to prevent overcharging of the battery further reduce the loss of water.

Brake Cylinder (or Master Cylinder)

BEHAVIOR

Allows you to use a small push with the toes to exert a huge braking action on the wheels. This is where hydraulic pressure is developed to operate the brakes.

HABITAT

The master brake cylinder is connected to the vacuum pump, which is mounted on the firewall.

HOW IT WORKS

The master cylinder has a reservoir of brake fluid to maintain the proper level of fluid in the system. Some have clear plastic components so you see the level of fluid inside without opening the lid. A low-fluid warning switch lights up on your dashboard if the level drops too much.

Most brakes today use hydraulics to increase the effective force. The force required to depress the brake pedal is amplified by a vacuum booster. The force then pushes a piston in the brake master cylinder. As the piston moves it compresses the hydraulic fluid, raising its pressure throughout the brake system. Since pressure is force per cross sectional area, the force varies with the size of the cylinder that contains it. The larger area of the wheel cylinders allows a huge force to be applied to the brake pad, which is needed to slow the car.

The larger the wheel cylinder is, the more force it can deliver with the same pressure in the brake line. However, the larger it is, the shorter distance it can move the brake pad. While your foot moves several inches in applying the brakes, the brake pad, housed in the wheel cylinder, moves but a fraction of an inch.

The master cylinder is split into two parts. Each part provides the braking power for two of the four wheels. So a leak or malfunction in one part still allows half of the brake system to operate.

Some trucks use air brakes instead of hydraulic brakes. Air brakes consume lots more space and use a compressor to provide the pressure in the system. When the air brake system is pressurized, the brakes are open and the truck can move. When the driver touches the brake pedal, pressure in the system drops allowing the brakes to engage. Trucks and trains use this system for safety: if a leak develops in the system, the pressure stops and the brakes engage.

INTERESTING FACTS

Brake fluid, like most fluids, is nearly incompressible: squeezing it doesn't change its volume appreciably. This trait, plus its high boiling point, is what engineers look for in finding liquids to use as brake fluid.

Brake fluid is categorized as Dot 2, Dot, 3, Dot 4, or Dot 5. Most cars in the United States use Dot 3, which is a polyethylene glycol–based fluid. Dot 5 is a silicon-based fluid.

Coil

Coils change the 12 volts coming from the battery into the 50,000 volts needed to power the spark plugs.

Under the hood, the coil is hung on the firewall. It has electrical connections to the battery and to the distributor.

A coil has two coils of wire inside. One, the primary, carries the 12 volts from the battery. The secondary coil has many times the number of windings as the primary. Current flowing through the primary coil induces a current in the secondary coil. When the current to the primary coil is suddenly switched off, a huge current is induced in the secondary coils. The relative number of windings in each coil determines the voltage induced, and since the second coil has many more its induced voltage is much higher. The high voltage in the secondary coils travels to the distributor before going to the spark plugs.

Dipstick

BEHAVIOR
It provides a reading of the engine's level of oil.

HABITAT
It is housed in an upward projecting metal tube on the side of the engine. The head of the dipstick is often a brightly colored plastic ring, making it easy to grab and extract.

HOW IT WORKS
The stick is a metal rod with markings etched on it so you can measure the amount of oil in the engine (crankcase sump). A similar dipstick is used to measure the level of transmission fluid in the transmission.

Distributor

BEHAVIOR
Serves up the electrical charge to each spark plug in its turn.

HABITAT
Distributors sit high above the engine. The top of the distributor has several plastic cylinders projecting upward, each with a wire leading out to a spark plug.

HOW IT WORKS
Inside the distributor is a rotating switch called the rotor. As it spins it delivers electrical pulses to the wires that power each of the spark plugs. It doesn't actually touch the contacts with the wires, but rides close enough to the contacts that a spark can fly from rotor to contact. Each spark plug is connected to one of the contacts.

The distributor also controls the flow of electricity in the coil. Since the coil has to deliver a blast of electricity at exactly the right moment to power the spark plugs, the switches that control the coil move with the spinning rotor.

New technology now makes distributor-less ignition possible. Using electronics instead of a mechanical system, current is distributed to spark plugs. This system removes several engine components that tend to fail over time and allows computer control over the distribution of electrical pulses.

Fan

BEHAVIOR
It spins quickly to draw air through the radiator to cool the fluid.

HABITAT
The fan sits directly behind the radiator, in front of the engine.

HOW IT WORKS
Fans are powered by either an electric motor or by a belt driven from the crankshaft. Engines in front-wheel-drive cars are mounted transversely, at an awkward angle to provide power to the fan by a belt. These have electric motors. These fans can be turned on and off automatically as the engine temperature rises and falls.

Engines in rear-wheel-drive cars have the crankshaft facing forward in position to drive the fan with a belt. The belt is a V-belt made of rubber with a steel reinforcement to prevent it from stretching. The narrow end of the V fits into the center of the pulleys. The belt driving the fan also provides power to the water pump.

The fan is most important when the car is idling or moving slowly. That's when the radiator needs additional air blowing on it to cool the engine. When the car is zipping down the highway, air flow through the grill provides enough cooling power.

Fan blades may number as few as two or as many as ten. They have rounded shapes and are designed to make the spinning blades less noisy.

Horn

It scares the daylights out of the unsuspecting pedestrian crossing in front of your car and possibly warns other drivers of preventable accidents. Where I live, in Seattle, it is considered uncouth to use one's car horn unless it's an emergency. However, I'm writing this while in the Caribbean country of St. Vincent where tooting a car horn is an essential part of driving a car.

HABITAT

The switch you depress to activate the horn is mounted on the steering wheel. The location of the switch is shown by the image of a horn. The noisemaker itself is located under the hood, usually close to the grill to give it unfettered access to the outside acoustic environment.

HOW IT WORKS

The car horn uses an electromagnet to flex a flat steel diaphragm. When you depress the horn button to get that slowpoke ahead of you moving, you send an electric current to the electromagnet. Now energized, this attracts a metal arm, which moves toward the magnet. As it does, it pulls the diaphragm with it.

That would be the quiet end of the story except that the arm now disconnects the circuit so the electromagnet loses power. The arm and diaphragm now flex back to their original position, which then closes the circuit again allowing current to flow. The electric current again moves the arm and diaphragm. As long as you have the horn button depressed, the diaphragm moves back and forth rapidly generating sound. The spiral horn or trumpet is sized to amplify the sound.

One side of the diaphragm is exposed to the atmosphere and as it flexes back and forth it vibrates air molecules. The energy of the vibrating air molecules travels at the speed of sound because it *is* sound.

The frequency of a car horn is typically in the range of 480 to 500 hertz, or cycles per second. The sound can be as loud as 110 decibels. To produce this much noise requires a lot of energy, which is supplied by the battery. Aside from the starter motor, horns require the largest current of any device in the car: about five to six amps.

The horn button doesn't directly activate the horn. To avoid having to use expensive, heavy wires to carry the high current, the horn button is part of a low voltage/current circuit that activates a relay switch that is located close to the horn. The relay switch operates the high current circuit that energizes the horn.

INTERESTING FACTS

Early cars had rubber bulb horns—giant versions of the toy horns on children's tricycles and bikes. In 1908 the Klaxon became popular. This horn has a center-stationary diaphragm. The diaphragm is moved by turning a crank or pushing an arm with teeth. The teeth catch and release the edge of the diaphragm to make sound. In 1911 electric-powered Klaxons became popular. The "ah-uaga" sound you hear in submarine movies when they are about to dive is a Klaxon.

Oil Filter

BEHAVIOR

Oil filters remove small bits of metal and other materials that accumulate inside an engine. Without a filter, the engine would need to have its oil changed every few hundred miles.

HABITAT

Oil filters screw into the engine block on the bottom or low on one side. Unless covered with oil and dirt, they are usually easy to spot

as they are often made of light or bright colors that stand out among the other dingy-colored components. The exception of this rule is the one shown here, which is black.

HOW IT WORKS

Oil is forced through the lubrication system by the oil pump. From there it flows through the filter where solid particles are removed. The oil then goes into the engine to lubricate the moving components.

The filter is often made of paper that has small openings that filter out the solids. As it captures more material, the passages through the paper fill up, restricting the flow of oil. Thus, the filter needs to be replaced periodically.

Most oil filters screw onto a threaded pipe coming from the engine block. They have to be seated properly and screwed firmly, but not too tightly, to ensure they don't leak. Some European and Asian car manufacturers use cartridge filters that fit into a pocket attached to the side of the engine.

Some people also use magnetic filters. Magnets catch and hold metal particles in the oil supply.

The first production oil filter was invented in the 1920s by Ernest Sweetland (#1,699,680). His patent claims the filter is useful for "clarifying the oil" and removing "deleterious matter from the lubricating oil." The filter medium was cloth supported on perforated plates. It had a viewing glass so you could see oil moving through the filter and would know when to change the filter.

Power Steering

This system makes steering a car much easier. Before power steering, professional drivers of cars and trucks developed muscular arms fighting to steer their vehicles. With power steering very little force is required to steer even a bus.

The visible parts of the system, not counting the steering wheel itself, are under the hood. The hydraulic fluid reservoir and pump can be seen just in front of the firewall of the engine compartment. This is where you should periodically check the fluid level using the dipstick inside the screw-off lid.

Beneath the car you can see the steering box. It pushes rods that move the steering arms connected to the front wheels.

Power steering is a hydraulic system. As you turn the steering wheel, its movement activates valves that admit hydraulic fluid under pressure. The power steering pump, located under the hood, provides the pressure. It is driven by a belt from the crankshaft.

From the pump, pressurized fluid travels in two hoses to the steering box. When the steering wheel is turned, the shaft from the wheel turns a torsion bar inside the steering box. The torsion bar turns the pinion gear of a rack and pinion steering system or a worm gear if it's not rack and pinion steering. As the torsion bar turns, it opens ports that let in pressurized hydraulic fluid that assists in making the turn.

Electric power steering is becoming popular. Sensors detect the amount of turning the driver applies to the steering wheel. A microprocessor then directs either a hydraulic pump or an electric motor to move the steering arms.

Electric systems have the advantage of using the car's power only when a turn is needed. Hydraulic systems consume power whenever the engine is running.

INTERESTING FACTS

Today most cars have power steering. With wider tires and heavier cars, steering without power assist would be difficult. Front-wheel-drive cars, which have more of their weight centered over the steering wheels are even harder to steer without power. The first car to come equipped with power steering was the 1951 Chrysler Imperial.

Radiator

BEHAVIOR

You may think of radiators as providing heat to a house on a cold winter day. But in cars, radiators take heat away— away from the engine and into the atmosphere.

HABITAT

Open the hood and you come face to face with the radiator. Of course, not all cars have them. Volkswagens, for example, are air cooled, not water

cooled, so they don't have radiators. But most cars have radiators and they are usually located directly behind the grill.

HOW IT WORKS

Radiators are devices that transfer heat from one place to another. Most radiators in cars use a mixture of water and antifreeze as the working fluid. This mixture is pumped by the water pump through the engine where it is heated by the combustive reaction of gasoline and oxygen in the air. The fluid continues to the radiator, where it circulates through tubes wedged between a honeycomb of metal slats. The slats provide a large surface area for cooling. As a car moves forward the air it encounters blows through the network of slats, picking up some of the heat. The fluid in the tubes is cooled by the passing air and it recirculates through the engine to remove more heat.

Gasoline engines convert the chemical energy in fuel to the mechanical energy that drives the car. But this conversion is only 30 percent efficient, and much of the energy released by burning fuel is lost as heat. Additionally, the motion of parts inside the engine generates more heat due to friction. The cooling system in general and the radiator in particular remove this heat. If the cooling system isn't functioning, the

RADIATOR CAP

RADIATOR OVERFLOW

engine temperature rises; parts expand and no longer fit together well, which generates more heat until the engine self-destructs.

The radiator cap sits atop the radiator. It provides pressure relief. The cooling system operates at high temperatures, and to prevent the coolant from boiling it is pressurized (raising the boiling point). The cap has a spring that forces a rubber gasket down to seal the radiator. If the pressure becomes too great, it can lift the gasket and fluid can flow into the plastic overflow tank adjacent to the radiator.

Spark Plug

BEHAVIOR

They create tiny bolts of lightening that ignite the fuel-air mixture inside the cylinders that causes the explosions that drive the car. They convert electric pulses into sparks that ignite.

HABITAT

Spark plugs are screwed into the engine head so their electrodes sit at the top of each cylinder.

HOW IT WORKS

Spark plugs are often covered by a rubber cap that protects them and the wire that brings electric charges to them. Looking at a plug itself, it is a cylindrical ceramic insulator inside of which resides an electrode.

An ignition coil or magneto creates the pulses of electricity that power the spark plugs. A large voltage difference, at least 20,000 volts and up to 100,000 volts, occurs between the two electrodes in the spark plug. The center electrode is the cathode or electron emitter. Across a small gap that separates the two electrodes, the voltage difference causes a spark. The gap needs to be the correct distance or the plug may not properly ignite the fuel in the cylinder.

Plugs fire about 15 times a second. Each firing occurs at a voltage of 40,000 to 100,000 volts.

INTERESTING FACTS

The same German inventor responsible for the automobile headlight, Gottlob Honold, invented the spark plug in 1902.

Starter

BEHAVIOR

The starter gets the pistons moving up and down in the cylinders and the valves opening and closing so the engine can operate. Gasoline and diesel engines, but not electric motors, require starters.

HABITAT

The starter is the cylindrical device that is found along the side of the engine, near the back. On top of the starter is a smaller cylinder housing the solenoid. The starter is connected to the flywheel with gears.

HOW IT WORKS

In internal combustion engines the pistons have to be moving before the engine can start. The valves also have to be opening and closing to let in the mixture of air and fuel and to allow exhaust gases to escape. It's the job of the starter to get the engine moving.

The starter is an electric motor powered by the battery. When the battery dies and you turn the key or push the starter button, you hear that low and slow *grrrr* sound rather than the normal fast rotation and engine starting.

When you turn the key you send electric current to the solenoid. The solenoid is an electromagnetic switch that turns on the powerful current needed by the starter to turn the engine.

The starter spins quickly and as it does a gear along its shaft is drawn in so it engages the flywheel. The flywheel is connected to the crankshaft, which controls the motion of the pistons. So the starter gets the pistons moving up and down.

As the engine catches, it rotates much faster than the starter motor. The faster rotation pushes the engaging gear outward so it disengages from the flywheel. Now the starter isn't connected and you can release the start button or ignition key.

FLYWHEEL

Before the self-starter was invented, starting a car was dangerous. The driver had to turn a hand crank at the front of the car. If the car back-fired, the crank could turn powerfully enough to break bones. At least one death was associated with starting a car.

Charles Kettering invented the self-starter in 1911, when he worked for DELCO. He obtained Patent #1,150,523. Several car makers turned Kettering down when he tried to sell them his new invention, but Cadillac used it in its 1912 models. The self-starter opened driving cars to a wider audience of people who previously couldn't or didn't want to turn the crank. The self-starter is one of some 140 patents Kettering obtained. He also was cofounder, with Alfred Sloan, of the Sloan Kettering Institute for Cancer Research.

Thermostat

BEHAVIOR

It allows the engine to warm up by blocking the flow of coolant until its temperature has reached its operating range. Once the engine is warm enough, it opens up, letting coolant enter the engine.

HABITAT

If you follow the hose leaving the top of the radiator back across the top of the engine, you'll end up at the thermostat. It is located on top of the engine.

HOW IT WORKS

Inside the thermostat is a wax that expands and contracts quickly as the temperature of the engine changes. When the engine is cold, the wax is a solid. As the engine warms, the wax melts and expands. It pushes up a rod on top of which a valve sits. The valve allows coolant to flow into the engine. When the engine is cool, the wax contracts and the valve drops back inside its housing (pellet).

Transmission

BEHAVIOR

Allows the car to travel forward and reverse while the engine only rotates in one direction, and allows the car to travel at widely different speeds even though the engine has a narrow range of rotational speeds.

HABITAT

The transmission is located directly behind the engine.

HOW IT WORKS

Internal combustion engines operate in a narrow band of speeds, but the wheels rotate over a much larger range and in two directions. To accommodate this mismatch transmissions were invented.

There are three types of transmissions. Manual transmissions are found in cars and trucks. A driver changes gears by depressing a clutch (to disengage the engine from the transmission) and then shifting gears with a gear shift lever, most often mounted in the floor to the right of the driver. Each position of the gear shift lever engages a pair of gears that provide a different gear ratio. Each ratio drives the wheels at a different speed for the same engine speed.

Most cars today have automatic transmissions. The driver moves the gear shift lever for forward (drive), reverse, or park. Once in forward, the transmission automatically selects the proper gear for the speed of the engine.

Instead of a clutch that connects the engine to the transmission, automatic transmissions have torque converters. These are hydraulic fluid–filled devices with an impeller and turbine. The impeller is connected to the crankshaft and is spun by the engine. The turbine is connected to the transmission. As the engine speeds up the impeller pushes fluid against the blades of the turbine and gets it to spin. A third component, a stator, controls how much torque is passed from impeller to turbine. At low engine speeds, the stator doesn't move, which increases the torque. At higher speeds it rotates with the impeller and turbine.

Continuously variable transmissions are found on golf carts, snowmobiles, and other small vehicles. Belts slide in and out on cone-shaped spindles delivering different speeds to the drive wheels.

Turbocharger

It adds the *vroom* to your engine. More specifically, it compresses air and forces it in the engine cylinders. With more air and more fuel in the cylinder the pistons can deliver higher horsepower.

The turbocharger sits on top of the engine. Its turbine wheel is exposed to the exhausting gases collected by the exhaust manifold. The other end of the turbocharger, the compressor wheel, connects to the air intake system between the air filter and intake valves.

Turbochargers are one type of superchargers. Superchargers in general compress air into the cylinders to get more bang for each stroke—a way of increasing an engine's power. Turbochargers and superchargers increase the air pressure entering the cylinder by up to 50 percent. Additional fuel is shot into the cylinder along with the air.

The distinction between turbochargers and superchargers is how they are powered. Superchargers get their energy directly from the engine's crankshaft. A belt connects the supercharger to a pulley on the crankshaft. Turbochargers are turbine-driven compressors powered by the engine's exhaust gases. As the gases leave the engine they spin the turbine blades. The turbine shares a shaft with the compressor that pumps air into the engine.

There are subtle differences between the performances of the two technologies. Turbochargers have an inherent lag between the time the driver steps on the accelerator and when the compressed air and fuel reach the cylinders. Both systems can radically boost a car's power (30 to 40 percent), but come at a cost. Engines have to be able to withstand the additional heat generated by the more intense explosions in the cylinders. The added heat can reduce the density of air entering the cylinder, opposing the purpose of the charger. To fix this problem, cooling devices are added to reduce the temperature of the incoming air. And the fuel mixture has to have high enough octane so the engine doesn't knock (experience premature explosions before the spark plug fires).

Superchargers operated at very high speeds, up to 150,000 rpm. At these speeds they make a distinctive, loud, whining sound.

INTERESTING FACTS

Working on steam engines in 1905, Swiss engineer Alfred Buchi came up with the idea of extracting energy from the exhaust gases of an internal combustion engine. Since a large percentage (up to two-thirds) of the total energy available in fuel is wasted in the exhaust gases, his idea was brilliant. The concept was adapted for use in train locomotives, cars and trucks, ships, and airplanes.

Water Pump

It pushes engine coolant throughout the cooling system to help keep the engine from overheating.

The water pump is secured to the front of the engine with bolts. Most are driven by a belt powered by the crankshaft.

The water pump is a centrifugal pump. It operates by rotating impeller blades that push the engine coolant around inside a cylinder. The impeller blades are usually rounded or scooped, but some are flat.

Due to the centrifugal acceleration of the cooling fluid, it moves to the outside of the cylinder where it can escape through an opening. The fluid then continues through tubing into the engine block, where it absorbs engine heat before it flows through the radiator and back to the pump.

The water pump is powered by the engine. A rubber belt connects it to the crankshaft of the engine. As the crankshaft rotates, the belt drives the water pump. The photo shows a water pump with the belt removed.

Windshield Cleaning System

BEHAVIOR

Allows drivers to wash the front windshield, and in some cars the rear windshield as well, with a push of the button.

HABITAT

The cleaning fluid is held in a reservoir found under the hood. The reservoir is made of polyethylene and its cap is marked to indicate what goes inside. The controls are often found on a stalk mounted on the steering column. Sometimes the rear cleaner control is mounted separately in the dashboard and the reservoir can be located in a side panel near the back of the car. Nozzles are mounted directly beneath the wipers and a pump is located under the hood.

HOW IT WORKS

An electric pump draws fluid from the reservoir and forces it out through the jets. Some cars have heaters to warm the fluid so it can melt snow and ice on the windshield.

Not that windshield cleaner fluid is very expensive, but you can make your own. One recipe is to mix 10 cups of water with 3 cups of isopropyl alcohol and add a tablespoon of liquid detergent. Shake well before serving to your car's reservoir. And make sure you're adding it to the correct container.

Windshield Wiper Motor

BEHAVIOR

It provides the muscle to move the wipers back and forth across the windshield.

HABITAT

The wiper motor is mounted on the top of the firewall at the back of the engine compartment.

HOW IT WORKS

This is a high-torque electric motor supplied with power from the battery. When you turn on the wipers, the motor converts electric energy into rotary mechanical energy. The output of the motor turns a worm gear, which looks like a spiral of metal wrapped around a metal rod.

Worm gears are fundamentally different from other gears in several ways. They can radically increase the turning power or torque, which is useful in applications such as windshield wipers. And they change the direction of rotation. In the windshield wiper the worm gear changes the direction of the motor shaft's rotation 90 degrees.

The worm gear meshes with another gear. This arrangement decreases the speed of the motor's rotation and increases again the torque. This second gear shares its axle with a cam or crank. These devices change the motion from rotary motion produced by the motor to the back-and-forth motion of the wipers. The cam or crank connects to a rod that drives the driver's wiper and to another rod that drives the passenger's side wiper.

The wiper motor is a direct current motor; its speed is governed by the voltage it receives. So to change the speed of the wipers, different voltages (up to 13 volts maximum) are fed to the motors.

Wiper motors have a park feature. When you turn the wipers off, they continue wiping, but stop in their normal rest or park position. To get the wipers to return to their park position when you shut them off, the motor has two additional electrical contacts. A circuit feeds power to them after you have cut power by switching off the wipers.

INTERESTING FACTS

Before electric motors were used to drive windshield wipers, the engine's vacuum pressure supplied the power. This meant that the wipers went faster when the engine went faster. Going up a steep hill reduced the speed of the windshield wipers. Electric motors replaced the vacuum pressure wipers starting in 1926.

OFF-THE-ROAD
PASSENGER VEHICLES

WHY STICK TO THE ROADS? Why stick to dry land? You can drive anywhere in an off-road vehicle. That's not to say you should drive anywhere, but vehicles have been designed for all types of driving environments. Cross a swamp, Arctic tundra, or the local lake—all are possible with off-road vehicles. Some are practical solutions to real problems and others are just fun.

Amphicar and Aquada

BEHAVIOR

It's a car. It's a boat. No, it's both! It drives on land and water.

HABITAT

Mostly found now in Amphicar shows or rallies organized by enthusiasts, you rarely get to see them on the road.

HOW IT WORKS

A 43 HP Triumph motor powers both the wheels and two small propellers that protrude from the rear end. This rear-mounted motor gives the amphibious car a top land speed of 70 mph and top water speed of 8 mph. It has no rudder and steers by the driver turning the front wheels.

The car is watertight so the occupants and their luggage are kept dry. But just in case, it does have a bilge pump. As a motorized boat and car, an Amphicar needs to be licensed for both. Not practical for most driving or boating applications, but in some cases it is an ideal compromise vehicle. Fewer than 4,000 were ever produced, all between 1962 and 1967 in Berlin.

Amphicars made some significant ocean crossings: from Africa to Europe and from England to France. The Amphicar was not the first automotive amphibian and not the last. Very recently a UK company has developed high speed amphibious technology and is selling amphibious cars called Aquada.

Instead of propellers, the Aquada uses a water jet for propulsion. An engine spins impeller blades that accelerate water and push it out the rear of the car/boat. Steering is accomplished through a nozzle that can swivel. To go in reverse, the impeller spins in the opposite direction. Besides going much faster across the lake than an Amphicar (as fast

as 30 mph), the water jet is safer since it has no external blades that spin.

The Aquada also has retractable wheels to reduce the water drag. Pushing one button retracts the wheels and disconnects them from the engine. When you come to the shore, drop the wheels and drive home. If you buy one, please give me a ride.

INTERESTING FACTS
Purchased new from 1962 and 1967, Amphicars cost less than $3,500. In 2006 a used Amphicar was sold for over $100,000.

All-Terrain Vehicle (ATV)

BEHAVIOR
ATVs make lots of noise while moving over rough terrain at amazingly fast speeds.

HABITAT
ATVs are found in rural areas, often on farms and ranches.

HOW IT WORKS
Most ATVs today are four-wheel models ridden by a single driver. Earlier three-wheel models were prone to rolling backward due to the light weight over the front wheel.

The driver steers with handlebars and controls the throttle by a hand-operated grip. The engines are similar to motorcycle engines and range in size from 50 to 950 cubic centimeters. They can be either two-stroke or four.

Transmissions can be manual, with up to five forward gears and a reverse, or continuously variable transmission that uses belts to change effective gear ratios. Most use either a shaft or chain to deliver power to the rear, driving axle.

ATVs allow drivers to venture into all types of terrain with large, low-pressure tires spreading the vehicle's and driver's weight over a large area. They have suspension systems (springs and shock absorbers) to remove some of the bumps and bruises on uneven terrain, but driving one is a dynamic exercise in shifting one's weight from side to side to maintain balance. With a vehicle weight of 250 to 500 pounds, the driver has to work to keep it under control.

INTERESTING FACTS
The first car that Henry Ford built in 1896 was an ATV predecessor called a quadricycle. It rode on four bike tires and the design was probably inspired by four-wheel bicycles (also called quadricycles) of the era. With only two forward gears, Ford's quadricycle could travel up to 20 miles per hour.

DUKW

BEHAVIOR

It carries people on land and water. A bus that floats.

HABITAT

DUKWs are found giving city tours in cities like Seattle and Boston that have prominent and navigable waterways.

HOW IT WORKS

The DUKW is a boat hull with six-wheel drive and a single propeller. It can travel at almost 6 knots on water and 50 mph on roads. A bilge pump spits out the water that seeps in.

The original design was dictated by the need to move soldiers and equipment ashore for amphibious assaults during World War II. Since driving conditions on land could vary from soft beach sand to hard pavement, DUKWs were outfitted with a device that let the driver change the air pressure in the tires: soft for sand and hard for roads. A two-cylinder air compressor fed a storage tank that the driver could draw on to increase the pressure of the tires. They were the first vehicles to use this technology.

It must have taken some creative minds to come up with the name DUKW and still comply with military nomenclature. The D indicates the year they entered service, 1942, and the U indicates that it was a utility craft, one that could be used for a variety of purposes. K refers to its front-wheel drive, and W means that both rear axles also have power.

INTERESTING FACTS

Some 21,000 DUKWs were built and used by the military from 1942 to 1945. After the war they were used for beach surveys and research and for rescue operations. The DUKWs seen today are mostly modern versions of the original craft and these carry sunscreen-slathered tourists on waterfront assaults.

Golf Cart

BEHAVIOR
Carries half of a foursome around 18 holes and to the 19th hole as well. It provides a speedy and effortless way to play a round of golf.

HABITAT
You find these fun-driving machines on golf courses and in retirement communities, especially in Arizona and California.

HOW IT WORKS
Electric motors or gasoline engines propel this buggy up to 15 miles per hour. Low-pressure tires spread the weight of the cart and cargo so not to damage grass. Even so, it is against the rules to drive on a putting green.

Golf carts have automatic transmissions that change the effective gear ratio from about 3:1 at low speeds to 1:1 at high speeds. The clutch has two parts: one attached to the crankshaft and the other to the shaft that powers the differential and turns the wheels. They are connected by a drive belt. The two clutches change in opposite ways as the engine speeds up. The one attached to the engine forces the belt to ride on a larger diameter shaft, and the one that powers the wheels moves onto a lower diameter shaft. At low engine rpms the engine turns three times for every revolution of the wheels, providing power. At higher engine rpms the engine turns once for each revolution of the wheels to provide speed. This magic is made possible with ingenious centrifugal weights and springs.

Gasoline engine golf carts have governors to limit the speed of the engine. The governor is there to prevent damage to the engine and not to limit your speedy driving. Some carts have a spring device that limits how far the accelerator can be pushed. Others limit the throttle by means of a device connected to the clutch that powers the drive wheels. A third type controls the spark plugs electronically.

Snowcat

BEHAVIOR
Drives nimbly over snow and other soft surfaces that would defeat a wheeled vehicle.

HABITAT
Seen grooming ski slopes and hauling people and equipment in polar regions. A mainstay for ground-based polar research.

HOW IT WORKS
There are many variations of the basic design. Tucker Sno-Cats use four sets of tread, while other manufacturers use two. The front two tracks provide steering on the Tuckers. Two-belt models typically use brakes to steer: the driver applies the brake to the side that he wants to turn toward.

Wide rubber belts are mounted on a fiberglass housing or on a series of pulleys that guide and tension the belt. The belts spread the load (upward of a couple of tons) over a large area so the pressure exerted on the surface is less than a pound per square inch. A drive sprocket coming from the transmission drives the belts.

INTERESTING FACTS
When Sir Vivian Fuchs led the first expedition to cross Antarctica (1957–58), he used four Tucker Sno-Cats. The machines ran at temperatures down to −70° F and gobbled gas at 1.5 miles per gallon while hauling sleds behind them.

The first motorized sled was patented in 1916 by Ray H. Muscott, #1188981. Muscott's model used skis on the front and tracks on the back.

Snowmobile

BEHAVIOR

Snowmobiles scamper over the snow and ice at incredible speeds. This is a personal transporter for cold conditions that long predates the Segway.

HABITAT

Seen in the front yards of many Alaskans and people living in the upper Midwest. As I write this I'm in Lulea, Sweden, where people are zooming down the frozen Lulea River on snow machines.

HOW IT WORKS

Two skies up front provide the steering and support while one or two rubber tacks in the rear drive the snowmobile. Lighter weight snowmobiles use a two-stroke gasoline engine, like the ones in lawnmowers. Larger ones use four-stroke engines.

The engine is mounted transversely. The drive shaft turns a large pulley. At low engine speeds a spring disengages the pulley, which serves as a clutch. As engine speed increases the centrifugal force closes the clutch so the pulley turns the drive belt. A secondary clutch connects to the drive axle and holds the belt during the initial movement of the snowmobile when it needs the most torque to accelerate. Once up to speed this secondary clutch disengages. This is a type of continuously variable transmission that can deliver a variety of speeds without noticeably shifting gears.

Power is delivered to the tracks by a driving gear. The teeth on the gear engage openings in the rubber belts. The belts are supported and tensioned by a series of pulleys and wheels.

You steer a snowmobile by turning the handlebars. The handlebars are connected to the two skis up front.

Snowmobiles float on the surface of the snow. They spread their weight and the weight of the driver over a large area, like snowshoes do for the walker. The tracks are made of rubber or compounds of Kevlar.

These machines are remarkably fast. Some go as fast as 100 mph. All are noisy and a nuisance to people seeking the solitude of nature.

INTERESTING FACTS

Industry pioneer Bombardier had intended that their new product carry the name of Ski-dog. They thought their machine would replace dog sleds. And to a large degree it has. But the painter made a mistake and they ended up with Ski-doo just before its unveiling, and without time to correct the mistake the name stuck.

HUMAN-POWERED VEHICLES

BICYCLES AND OTHER human-powered vehicles provide hours of enjoyment on a sunny spring day and a practical means of transportation for those energetic enough to push them along the road. Typically, they hide in garages and sheds across America. More are seen there than out where they belong on roads and bike paths.

As simple as the machinery is, the physics are complex. On a bike, for example, the rider balances weight and adjusts the front wheel position to maintain balance. Riding in a straight line requires balancing over the centerline of the bike. Turning a corner requires the rider to lean into the turn—or face the consequences of a fall toward the outside. The lean is required to counteract the bike's tendency to continue traveling in a straight line. With the front wheel turned and the rider sitting vertically, bike and rider will topple in the opposite direction of the intended turn.

Nearly all bikes transfer the power of the rider's quadriceps to the rear wheel by means of a chain. A derailleur system moves the chain toward and away from the bike so it can mesh with different gears. A spring-mounted sprocket tensions the chain and allows its effective length to change so it can connect different-sized gears. Large gears on the front sprocket (near the pedals) drive the bike fast when connected to small sprockets on the rear wheel.

Braking is achieved by hand pulling two levers, one for each wheel. Depressing the levers pinches the rim of the wheels with rubber brake shoes.

Most other human-powered vehicles are driven by the leg muscle as bikes are. Many have adapted bicycle technology for braking and turning.

Kirpartrick MacMillian of Scotland invented the first mechanical bicycle in 1839. He was trying to build a device that would allow him to travel longer distances than he could cover on foot. He wanted to visit his relatives who lived 40 miles away. His invention didn't catch on and new efforts later in the century led to the first U.S. patent for a bike in 1866. Innovations followed including hollow tubes instead of solid metal bars, forks to hold the front wheel, and later pneumatic tires. A bicycle craze hit America and people socialized at indoor riding halls, streets and roads being too rough to ride on.

In 1870 Britain's James Starley invented the penny-farthing, the odd-looking bike with a huge front wheel and small rear wheel. Starley's nephew, John Kemp Starley, introduced the safety bicycle in 1884. The safety had a chain-driven rear wheel and a triangular frame. Many modifications later, the safety bike is the basis for the bikes we ride today.

Bicycle Escalator

BEHAVIOR
It propels you and your bicycle to the top of a tall hill.

HABITAT
The only one is found in Trondheim, Norway.

HOW IT WORKS
Think of this as a ski lift without skis or snow. Or an escalator without steps. A large electric motor pulls an endless wire rope that is supported at each end on pulleys that are about 24 inches in diameter. Attached to the rope are 11 evenly spaced foot plates. One of the foot plates cycles past every 12 seconds when the motor is running. The motor turns off when no one is using the system.

To help accelerate the mass of bike and rider from a stop at the bottom of the hill, a piston-action accelerator launches the rider at the start. Then the wire rope provides the lift to the top. Up to six riders can use the system at one time. Some parents even use the escalator to push baby carriages up the steep hill.

The goal of the operator is to make bike riding and bike commuting more popular. Bike commuters can navigate this hill, which is in the third largest Norwegian city, without working up a sweat, so they can show up at work and not have to change clothes.

Tourists can borrow a key from a nearby bakery/cafe to operate the lift at no cost. Residents can purchase an annual pass key for about $16.

INTERESTING FACTS
Between 20,000 and 30,000 people each year catch a lift on the bike escalator. It opened in 1994.

The city of Trondheim also offers free bicycles. Anyone can get a bike pass from the chamber of commerce. The pass opens bike stall locks throughout the city where bikes are kept. Open the bike lock and ride off. When you are finished, return the bike to any of the locks.

Bicycle Suspension System

BEHAVIOR

Makes riding over rough terrain possible by absorbing the vertical motion resulting from hitting bumps and holes in the road. Without a suspension system, riding off-road is jarring, uncomfortable, slow, and dangerous.

HABITAT

Suspension systems are found on the front fork and supporting the rear wheel or supporting the seat.

HOW IT WORKS

Suspension systems are combinations of large springs and hydraulic shock absorbers. Pushing oil through the openings of the shock absorber converts the energy of motion into heat.

The bumps encountered by the front wheel are absorbed by two springs made of steel, titanium, or compressed air. The springs are contained inside cylinders that support each fork, or the springs can be exposed on each fork. Springs absorb much of the vertical motion, but without dampening they would continue to bounce up and down.

A hydraulic shock absorber slows down the bouncing motion of the springs. Inside the shock absorber each motion forces oil to move through small openings as the wheel moves up and down. The moving oil transforms some of the motion energy into heat and reduces the bouncing.

Seats can be supported by an elastic support that absorbs shocks. The rider sits at the end of an arm that can flex up and down as the bike moves.

The rear suspension often has a large, single spring and shock absorber. The wheel can pivot around a pivot attachment on the bike frame. The suspension connects the wheel to a different point on the frame and reduces the motion around the pivot point.

Brakes

BEHAVIOR

They slow down and stop a bicycle and prevent many accidents. They convert the energy of motion, laboriously added by your pumping legs, into heat.

HABITAT

Brakes are found on both front and rear wheels. Most commonly they ride near the upper side of the rim of each wheel and connect to hand controls on the handle bars.

HOW IT WORKS

Bike brakes come in three designs. Rim brakes are the most common. Squeezing the hand lever forces two brake pads made of rubber to rub against the metal rim of a wheel. The hand lever pulls a cable that draws together two arms that straddle the wheel and rotate through a pivot point. Alternatively, the cable can pull on one side (single pivot side-pull) to draw the two arms together.

Disc brakes are found on mountain bikes. A metal disc is attached to the wheel hub and spins between the two sides of a caliper brake. The rider activates the caliper with a hand lever and the action is conveyed to the calipers either by a cable (mechanical system) or hydraulic system. The hand lever forces hydraulic fluid through a tube that forces the two sides of the calipers to come together.

Most bikes used to have coaster brakes, and a few still have them. Coaster brakes slow the rear wheel when the rider pedals backward. The brake is inside the rear wheel hub. Pedaling forward disengages a clutch, which lets the wheel spin. Pedaling backward forces a pin into the clutch, which pushes outward against brake pads.

Or, when all else fails, take the gloved hand and press it against the front tire. Or drag your shoe along the ground. Eventually you will stop.

Derailleur

BEHAVIOR
This nifty device allows bike riders to change gears. It is the manual transmission for bicycles.

HABITAT
Most bikes today have two derailleurs: one mounted up front, near the cranks, and the other one mounted near the rear axle. The chain runs around each.

HOW IT WORKS
The rear derailleur is more complex, as it has two functions to perform. A small, toothed wheel in front and below of the rear axle provides tension on the chain. It is held in place by a spring so it can take up slack as you change gears. The smallest rear gear, which provides the highest speeds, requires less chain than does the largest rear gear—"granny gear" for going up hills. To accommodate the differences in the size of the gears, the tensioner pushes down on the chain. It pushes on the bottom of the chain where there is little tension. The top of the chain carries the mighty pushes of your quadriceps from front gear to rear, but the bottom of the chain hangs out, loose and easy.

The rear derailleur also moves the chain in (toward the bike) and out (away from the bike) so the chain coming up from below will land on the gear you want. A hand lever on the handlebars or stem pulls a

cable that draws the derailleur in and out against the restoring force of a spring. Rear derailleurs typically serve five or more gears.

Since the chain comes to the front derailleur on the top of its route, it is under tension on this leg. No tensioning device is needed. The front derailleur moves in and out like the rear derailleur does, with its control mounted on the opposite side of the handlebars (usually the left side).

There are two types of shift mechanisms used with derailleurs. In the more traditional one the rider pushes on the gear shift lever until she hears and feels the chain has moved onto the next gear. Then she fine-tunes the positioning to eliminate any clicking or grinding noises that indicates that the chain is not centered on the sprockets.

Indexed systems move the cable into the correct or centered position for each gear. The rider just pushes the lever until she hears and feels it click into position.

There are many variations on this basic design and new designs continue to appear.

INTERESTING FACTS
Derailleurs have been used for over a century, but were not common until cable-derailleurs were introduced in the 1930s.

Quick-Release Hub

BEHAVIOR
Allows riders to unlatch front and rear wheels in seconds instead of digging to find the correct wrench and having to take off the nuts that hold the wheels onto the axles.

HABITAT
Found very often on front wheels and increasingly on rear wheels as well. The lever arm protrudes from one end of the axle.

HOW IT WORKS
The lever arm pushes a cam that provides the tension to hold the wheel on the axle. A threaded rod passing through the axle holds a nut on one end and the quick release lever on the other. The nut is screwed onto the rod, which pushes the forks against the axle. When the nut is screwed on far enough, the lever on the other end of the rod is rotated 180 degrees, and this pushes a cam that provides the final tension to hold the wheel in place.

Quick-release systems (quick-release skewers) are also used to hold the seat post in the bike frame. This allows the rider to adjust the height of the seat without using wrenches.

INTERESTING FACTS
The idea for the quick release mechanism came from Gentullio (Tullio) Campagnolo, an Italian bike racer and bike innovator. He invented improved derailleurs, brakes, and pedals in addition to inventing the quick release.

Pedicab or Cycle Rickshaw

BEHAVIOR
Pedicabs are human-powered taxis. One person powers the cycle and one or two people ride.

HABITAT
These bicycle taxis are used in Asia, Europe, and North America. In some places they are an economical alternative to motorized cabs. In resort towns, they provide a fun ride, worthy of a vacation photograph.

HOW IT WORKS
There are many varieties of cycle rickshaw. Most common are three-wheeled cycles where the driver pulls the riders. But in some countries the cyclist rides behind the cab. Sidecars are another version. Other pedicabs have four wheels.

In all, the cyclist pedals the cranks that turn a chain to power either one wheel or an axle with two wheels. Some have either electric motors or gasoline engines to assist the cyclist.

INTERESTING FACTS
Human-powered rickshaws are not that old. They were invented by an American in the late 19th century. Although the story is disputed, apparently an American missionary living in Japan built a rickshaw so his invalid wife could get out of the house.

Unicycle

Provides a one-wheel vehicle for those who are well-balanced. An inefficient but fun method of travel, one that attracts attention.

HABITAT
Found in parades, festivals, and downtown malls. Increasingly they are found on back country trails.

HOW IT WORKS
Similar to a bicycle, a unicycle has a bike wheel with pedals and cranks, a fork that holds the wheel, a seat post, and seat. Unlike a bike, most unicycles have no gears, brakes, handlebars, or second wheel. With no gears, the top speed of a unicycle is directly related to the size of the wheel and how fast the rider can move his or her legs.

Dynamic balance is key to riding the unicycle. It represents an inverted pendulum, with the weight at the top (the rider) and the balance point at the ground. The rider maintains balance by pushing the pedals forward and backward to keep the wheel beneath. The rider turns side to side to maintain balance along that axis.

Although not easy, street riding is relatively simple compared to stunt riding and mountain trail riding. Riders bounce up and down to hop logs, and although it is easy to lose one's balance, it's easy and safe to step off a unicycle to avoid a fall.

Some unicycles have a brake, which comes in handy for slowing down on a long downhill. The brake control is located under the seat. Some also have gears. Usually there are just two gears: 1:1 and 3:2 gear ratios.

INTERESTING FACTS
The Web site www.unicycling.org lists several unicycle records. Imagine riding a unicycle that has a wheel of 73-inch diameter. Long

legs come in handy. Or imagine riding one that is 115 feet tall. No thank you. Ever heard of a two-wheeled unicycle? The record for most wheels on a unicycle is apparently 13. Only one wheel touches the ground; the others act as gears conveying the rider's power to the one driving wheel. Image a unicycle built for two people. These unicyclists are a pretty creative bunch.

Kick Sled

BEHAVIOR

Kick sleds provide a faster-than-walking means of transportation across snow and frozen surfaces. They are the equivalent of scooters for ice and snow.

HABITAT

Kick sleds are found in Norway, Finland, and Sweden and in other countries with Scandinavian pop-

ulations. People use them where people in warmer climates would use bicycles to commute to school or to go shopping. They are found locked (with bike locks) outside stores in northern Scandinavia.

HOW IT WORKS

The driver stands on one runner with one foot and uses the other foot to kick to the rear to propel the sled forward. The runners are thin and flimsy steel that flex as they encounter hard ice or rocks. By twisting the handlebars, the driver twists the runners to change directions. Although thin, their length gives them the surface area to support the weight of the driver and groceries or a passenger.

INTERESTING FACTS

Although people have been using sleds in northern climates for centuries, the kick sled appears to be a recent development. The availability of iron or steel (Sweden has the largest in-ground iron mine in the world) in the late 19th century probably led to its development. Originally the runners were rigid, but around 1900 the design changed to the flexible runners to allow better turning.

Scooter

BEHAVIOR

Scooters provide an environmentally friendly method of transportation at low speed.

HABITAT

Scooters are found in garages of homes across America where the young or young-at-heart live. And they are found working in factories and airports where workers have long distances to cover and materials to haul.

HOW IT WORKS

Some fold up for easy storage, but the scooters that are used for moving packages and bundles have rigid frames. The driver stands on the scooter with one foot and pushes or kicks the ground backward with the other.

Steering is accomplished by twisting handlebars. They are connected directly to the single front wheel. There can be either one or two rear wheels. Some scooters have package baskets on the rear or smaller baskets hanging from the handlebars.

These photos were taken in the Oslo airport, where many of the airport workers use scooters to restock retail stores and to navigate quickly through the crowds with little danger of collision.

MOTORCYCLES

VEHICLES WITH TWO, IN-LINE WHEELS powered by an engine are generally called motorcycles. There are many variations of these vehicles, but we will lump them together.

A steam-powered motorcycle may have existed before the gasoline-engine motorcycle was invented by the inventors of gas engines. Sylvester Howard Roper created a steam-powered motorcycle in 1867, years before gasoline engines. Roper had a heart attack and died at the age of 73 while driving his invention at a record-breaking 40 mph. Today almost all motorcycles are powered by gasoline engines. A few are electric and some are diesel.

Gasoline-engine motorcycles were invented in 1885 by the same two men who invented the gasoline engine, Gottlieb Daimler and Wilhelm Maybach. They mounted an internal combustion engine on a bicycle. Only later did they use their engine to power a wagon to create one of the first gasoline-powered cars.

Like cars, motorcycles run on gasoline. Unlike modern cars, many use carburetors to mix the fuel with air and send it into the cylinder(s). Motorcycles can have one or several cylinders, and by looking you can easily see how many cylinders there are. Each cylinder is housed in a steel jacket (engine case) with parallel ridges to help dissipate the heat. Some of the larger cycles are water-cooled, but most are cooled by contact with the passing air. An exhaust pipe exits each cylinder and ends in a muffler and tailpipe. If a bike has two or four cylinders it may have two exhaust pipes and mufflers, one on either side. A manifold collects the exhaust gases from two cylinders on the same side of the bike before sending the gas toward the rear and the muffler.

Power from the engine passes to the transmission. The driver operates a clutch, usually with the left hand. Squeezing the handle pulls open spring-controlled plates that allow the engine to turn without power going to the rear wheel. Releasing the grip allows the spring to close the clutch plates so power is transferred.

Most bikes have five or six forward gears and no reverse. The gears are accessed in sequence so the driver can't skip a gear. The gear shift lever is operated by the driver's left foot, although on older models it is operated by the right foot.

Most motorcycles use a chain drive from the transmission to a large sprocket mounted on the rear axle. Some however use a shaft drive, similar to the ones used in automobiles. And some use a belt drive.

The rear wheel connects to the bike frame by a swing arm that allows the rear wheel to pivot up and down. Springs and a shock absorber support the rear wheel and absorb the bumps of the road.

The front wheel is held by two forks into which the axle is bolted. To absorb bumps the forks can telescope in and out with shock absorbers mounted inside the forks. Brakes can be either disc or drum, just as in cars.

One advantage of motorcycles is that their low weight gives them good fuel economy. The big disadvantage is the unpleasant weather and driving conditions, and lack of physical protection in the case of collisions.

Brakes

BEHAVIOR

Brakes slow the motorcycle down. They convert mechanical energy of motion into heat that escapes to the air.

HABITAT

The front brake control is the large lever on the right side of the handlebar. The rear brake control is a foot pedal, usually also on the right side. The brakes themselves are mounted either in the wheel hubs or on large discs that attach to the wheel hub.

HOW IT WORKS

Each wheel has its own brake. Like in cars, motorcycles have either of two kinds of brakes: drum (top photo) and disc (middle photo). You can't see drum brakes, but can see the brake cable leading into the wheel hub. Inside the drum brake, pads are forced outward to press against the spinning drum to slow the motorcycle.

With disc brakes you can see the large, often silver-colored, brake disc. A brake caliper holds an opposing pair of brake pads that squeeze the brake disc to slow it down.

A reservoir for brake fluid is often mounted on the handlebars, usually on the right side.

Carburetor

BEHAVIOR

It mixes air and fuel and feeds the explosive mixture into the cylinder, ready for combustion.

HABITAT

On this motorcycle the carburetor sits behind and below the fuel tank, and immediately behind the cylinder.

HOW IT WORKS

The carburetor is essentially a pipe that carries air and introduces fuel to the engine. The pipe is constricted in the middle, which causes the entering air to speed up. At the narrowest part of the tube, small holes (jets) admit the gasoline, which is sucked in by the low pressure of the fast-moving air.

As the rider opens up the throttle, it admits more air through the air filter into the carburetor. The faster moving air draws up fuel (Bernoulli's principle) and carries it into the cylinder.

Downstream of the constriction, where the tube has widened again, is a butterfly valve or throttle valve. This is what the rider controls. With the throttle closed, the valve stops the flow of air through the tube. Fully open, the valve is rotated 90 degrees in the tube so it allows maximum flow of air (and fuel).

Upstream of the constriction is a choke valve. The rider can close the choke when starting a cold engine. It reduces the flow of air, which increases the ratio of fuel to air in the engine, making it easier to start. Once the engine has warmed up a bit, the choke is opened so the engine gets a normal ratio of fuel and air.

Larger and more expensive bikes come with fuel injection systems rather than carburetors, but most bikes on the road still use carburetors.

INTERESTING FACTS

The first carburetors used wicks to draw up the liquid fuel and allow it to mix with air. One of the early pioneers of gasoline engine vehicles, Wilhelm Maybach, invented the float-feed carburetor in 1893. Henry Ford patented an improved carburetor in 1898, Patent #610,040.

Engine

BEHAVIOR

This is where the *vroom, vroom* sounds start. Explosions of fuel mixed with air occur here to drive one or more pistons down inside the cylinder(s). The pistons transform the chemical energy in fuel to the mechanical energy that moves the bike.

HABITAT

The engine is bolted onto the frame and sits directly beneath the rider.

HOW IT WORKS

Nearly all motorcycle engines run on gasoline. Most have one or two cylinders, although some motorcycles have more—anywhere from three to eight.

Like other gasoline engines, motorcycle engines take a mixture of fuel and air into the cylinder(s). The piston compresses the mixture and, at the point of greatest compression, a spark plug ignites the mixture. The explosion drives the piston downward in the cylinder. The bottom of the piston is connected to a crankshaft. The cranks on the crankshaft convert the up-and-down motion of the piston into rotary motion.

Engines and the motorcycles they power are described by their total displacement in cubic centimeters (ccs). A very small bike might have a 50 cc displacement while a Hog might have 500 cc or more. Displacement is a measure of the total volume of all of the cylinders between the top of the pistons (when in its lowest position in the cylinder) to the top of the inside of the cylinder.

Motorcycle engines can be either two-stroke or four-stroke. Car engines are all four-stroke, which means that the piston moves up and down four times for each fuel explosion. Two-stroke engines are generally dirtier (emit more air pollutants) and cheaper, and deliver a rougher ride.

Most motorcycle engines are air-cooled. The outside of the engine case is lined with heat-dissipating ridges. Some are water-cooled, which requires a radiator to get rid of the heat. A few are cooled by oil. Engine oil is circulated inside the engine case and out to a radiator.

INTERESTING FACTS
Almost as soon as motorcycles were available, police departments started to use them. Berkeley, California, had the first recognized motorcycle patrol in 1911. Individual officers in other cities had been using motorcycles to patrol as early as 1908.

Exhaust System

BEHAVIOR

The exhaust pipes direct the exhaust gases from the cylinder to the back of the bike, through a muffler, and out into the atmosphere.

HABITAT

On fancy bikes, the exhaust pipes are chrome or silver in color. They are easy to see: the large pipes that start at the engine case and snake backward, ending near the top of the rear wheel.

HOW IT WORKS

Single-cylinder bikes have one exhaust pipe that feeds the muffler before venting to the outside. Bikes with two or more cylinders have one exhaust pipe on either side. Bikes with four or more cylinders have manifolds on each side to collect the exhaust gases from the cylinders on each side.

The bump you see near the end of the exhaust pipe is the muffler. It contains some baffles that block some of the sound—but often not enough of the sound.

Foot Controls

BEHAVIOR
Foot controls allow drivers to control the clutch and rear brake.

HABITAT
On this older British motorcycle the controls are on opposite sides from most bikes. Most bikes have the rear brake on the right side and the gear-shifter on the left side. Also on the right side is the starter pedal. Note that it folds up when not being used to kick-start the bike. Behind the foot controls are pegs for the driver to rest his or her feet.

HOW IT WORKS
The brake and gear shift levers pivot around an axle and connect to a mechanism beneath a metal cover. Push down on the forward gear shift lever to shift up and push down on the rear lever to downshift.

Gasoline Tank

BEHAVIOR

This is the reservoir for fuel. Notice how small it is compared to the fuel tank in a car.

HABITAT

It sits high atop the frame. The rider straddles the tank.

HOW IT WORKS

In a word: gravity. Gasoline flows from the tank into the carburetor and from there into the cylinder.

Getting gasoline into the tank isn't as easy as with a car or truck. The nozzles at gas stations are designed for these four-wheel vehicles, not for motorcycles. The gas fume recovery sleeve on the nozzle (the pleated rubber outer covering) has to be pulled back and held by hand to fill a motorcycle. And, unlike mindlessly filling a car, the motorcyclist must pay attention not to overfill the tank and spill gasoline all over the bike.

INTERESTING FACTS

A gas tank on a motorcycle is one of those items you'd think would never cause a problem. But rust can form on the inside of the tank and flecks of rusted metal can fall off and get sucked into the engine. Most motorcycle manufacturers do not coat the inside of their gas tanks so any water that accumulates in the tank can cause rust. Tanks also get pinhole leaks and cracks, which can be repaired by welding or soldering. Welding a tank that contains gas fumes is an activity only professional welders should attempt. (They empty the tank and fill it with an inert gas before welding).

Hand Controls

BEHAVIOR

The hand controls allow the driver to apply brakes (to the front wheel), sound the horn, apply the choke for cold starting, adjust the throttle, stop the engine, and control the clutch.

HABITAT

The right hand operates the front brake. This is the large lever adjacent to the right grip on the handlebar. The plastic grip itself is the throttle control. The driver controls the throttle by twisting the grip clockwise or counterclockwise. A small button on the handlebar allows the driver to kill the engine.

The left hand operates the clutch by the large lever adjacent to the left grip. A small button for the horn sits where the left thumb can access it. Another control mounted on the handlebars farther inboard is the choke.

HOW IT WORKS

The controls on most bikes are operated by cables that connect to the controls on the handlebar and to foot controls.

Oil Tank

BEHAVIOR
It holds engine oil.

HABITAT
The oil tank is usually mounted above the engine.

HOW IT WORKS
Engine oil is contained either by a visible tank such as the one shown here or in the bike frame itself. Oil-in-frame chassis bikes use open space within the frame instead of a tank.

The job of the oil tank, besides being a reservoir, is to mix the oil. The intake and outtake oil lines are located so they circulate the oil in the tank.

INTERESTING FACTS
Some motorcycles use oil to help cool the engine. The oil is circulated between the engine and a small external radiator. These bikes are called "oil-cooled."

Oil tanks and gasoline tanks on motorcycles can be works of metallic art. Custom-made tanks adorn both show bikes and street bikes.

Radiator

BEHAVIOR

It cools the coolant that flows through the engine to maintain ideal operating temperatures.

HABITAT

It sits at the front of the frame, in front of the engine and behind the front wheel.

HOW IT WORKS

Water or antifreeze coolant circulates between the radiator and the engine. In the engine the fluid picks up heat by conduction. In the radiator it gives off the heat to the air.

Although most motorcycles are air-cooled, some are water-cooled. A third variation is to use the engine oil as a coolant.

Shock Absorbers

BEHAVIOR
Shock absorbers remove some of the bumps in the road.

HABITAT
Shock absorbers are built into the fork for the front wheel. For the rear wheel the shocks connect the wheel assembly to the frame behind the driver.

HOW IT WORKS
The front wheel is held to the frame between the two sides of the fork. The fork connects to the wheel axle at its lower end and to the frame at its upper end. Each side of the fork telescopes—that is, each side has an upper and lower component. One part slides inside the other. Inside the forks are springs and hydraulic shock absorbers. The springs slow down the rapid compression and expansion of the forks as the bike hits potholes and speed bumps. The hydraulic shock absorbers dampen the up-and-down motion of the springs.

Shocks for the rear wheel can be mounted on each side, as the one shown here. This is an adjustable shock that the driver can change for different road conditions. Rear shocks have a spring mounted over the shocks. Some bikes have a single shock absorber for the rear wheel.

Sidecar

BEHAVIOR

Sidecars attract attention wherever they go. They provide a second seat on a motorcycle.

HABITAT

Sidecars are found on the left side of a small number of motorcycles.

HOW IT WORKS

Sidecars transform motorcycles from two-wheel vehicles to three-wheel vehicles. Superficially similar in appearance to motor-tricycles, sidecars are fundamentally different.

Sidecars initially were intended to be easily removable so a motor-cyclist could quickly change the configuration. In practice, they become permanent attachments. The sidecars attach to the frame of the motorcycle and are supported by the motorcycle on one side and their wheel on the other. The sidecar's wheel is not connected to the motorcycle's drive train and is usually not in line with the motorcycle's rear wheel. Motor-tricycles, in contrast, have two rear-drive wheels that are aligned.

Sidecars were used by the German military in World War II and continue to be used in racing. It is a rare treat to see them on the street.

Segway

BEHAVIOR

This personal transporter carries one person up to 12 miles per hour on sidewalks and streets. It is a self-balancing, two-wheeled vehicle.

HABITAT

It can be found under the feet of a few thousand people who are rich and hip or who work for delivery companies or governmental agencies.

HOW IT WORKS

The Segway is a marvel of engineering. Powered by a rechargeable battery, it senses what direction you want to go and moves in that direction.

Gyroscopic sensors, not magic, tell it where you want to travel. Leaning forward, like you would to start walking, starts you moving forward. MEMS (micro electro-mechanical systems) gyroscopes detect your leans. The MEMS gyros have a tiny silicon plate that is energized by electrostatic charges, which causes silicon particles to vibrate. When the plate is moved because the rider leaned in one direction, the particles change their motion. This change is detected and guides a computer to control the two drive motors. Segway has five of these MEMS sensors. Sensing is done 100 times per second so the Segway can maintain balance on just two wheels.

Ten microprocessors interpret the sensor input and control the motors. A governor limits the speed of the Segway by forcing the rider to lean backward at top speed. Otherwise a rider could lean farther forward than the machine would be able to compensate for and would fall forward.

The concept is that you lean forward and the Segway moves forward so you always stay balanced on top of its center of gravity. It's just like walking, where you lean forward and thrust out one leg to catch you. There are no brakes; if you want to stop while moving forward, you lean back.

At top speed you can travel for an hour to an hour and a half before needing to recharge the batteries. If you need to put it in the back of your pickup truck, start eating your Wheaties—Segways weigh a bit over 80 pounds.

INTERESTING FACTS

Dean Kamen invented the Segway. Although it is a remarkable machine, its sales have not lived up to the hype surrounding its launch. About 35,000 have been sold, but predictions were for many hundreds of thousands. Of course, sales might yet take off.

RAMP

**Kneeling
Bus**

9 BUSES

STAGECOACHES WERE FIRST BUILT in the United States in 1827. Before that people traveled mostly by foot, on horseback, or by boat. The introduction of motorized buses and trolleys led to the demise of horse-drawn stagecoaches.

About the same time that stagecoaches were being first built in the United States, France and Great Britain initiated the first public transit system. The word *bus* is shortened from the original name, *omnibus*. Electric trolley buses got started in Germany in 1882.

Today there are many kinds of buses, from the big yellow Bluebirds to the articulated city buses and trolleys. Most are powered by diesel engines like diesel cars, but much larger. They have some interesting technology that you can see as you walk on board or as a bus zips by.

Bus Tracking System

BEHAVIOR

Allows dispatchers to monitor the location of all the buses and to immediately see if any bus has a problem.

HABITAT

The antenna is mounted on the front of the bus. The electronics are inside the bus.

HOW IT WORKS

Older systems use a variety of technologies to track their buses. One system relies on the bus's odometer and battery-powered signal radio transmitters mounted on road signs. The data is sent by radio to a computer at a central office.

In the operations center screens show the location of each bus on a map. The driver of each bus can be identified by his or her personal code, which is entered when starting the bus. If the driver has an emergency, he or she can depress a button that causes the bus locator to flash on the map. A dispatcher can notify police or other emergency responders.

The system can also control traffic lights. Each bus has an RFID tag on the front. Some intersections have readers that will change the signal timing to favor a bus.

Newer systems rely on the global positioning system (GPS). Each bus has a GPS receiver that receives satellite signals and computes the bus's location every few seconds. This information is relayed to the operations center either by a cell telephone network or by a dedicated UHF (ultra-high frequency band) radio link.

Fare Box

BEHAVIOR
It collects your money and puts it into a safe box.

HABITAT
As you are walking up the stairs of a bus, it meets you at the top.

HOW IT WORKS
Gone are the days when you paid cash to the driver, who made change for you. Today you insert exact change or flash your RFID card in front of the reader.

The Orca device in the photograph is an RFID card reader. It is a radio frequency identification system. The rider carries a smart card that can identify the rider and report the balance of her account. When in close proximity to the fare machine, the rider's card responds to a radio signal prompt and sends its identification and balance to the fare machine. The cost of the fare is then deducted from the account. When the bus is back in the yard, the card reader communicates by Wi-Fi with the accounting office to update its records. Riders can add funds to their accounts online. The next time they use the RFID card reader, it will update their account on their RFID card.

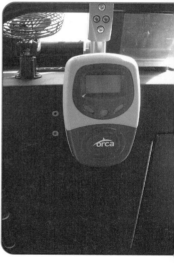

A dollar-bill reader and change counter are included in the fare box. To ensure that the piece of paper you are stuffing in the box is real, optical and magnetic sensors examine the bill. If the currency detector approves the bill, it reads the value and sends that information onto a

microprocessor that signals the driver that you've paid and records the amount. Coins are sorted by size and weight and are checked for their magnetic properties and their optical appearance.

The cash falls into a secure vault below. The driver doesn't have access to the vault and never sees or counts the money received. At the end of the day the driver parks the bus in a bus yard. An equipment service worker removes the money as well as refuels the bus and cleans it. But even this worker doesn't touch or even see the money. He removes the sealed vault and inserts it into a collecting device that opens it and removes the money. Once empty, the worker returns the locked vault to the fare pedestal. The fare box can send its accounting data to an office by Wi-Fi when the bus is in the yard.

INTERESTING FACTS

Before the fare box, fares were collected by a conductor. He walked throughout the bus or stood at one doorway collecting money and making change. To keep track of the collected fares, the conductor would pull on the overhead wire that was connected to a register. Of course, if the conductor didn't register all the fares, he could keep some for himself, so bus operators were interested in having the fares collected without human contact. Tom Loftin Johnson invented the fare box in 1880.

Outside the Bus

Notice the hinged small flaps around the back of the bus? These give access to equipment service workers to check or fill various fluids. There is one for oil and another for engine coolant. The fuel flap lifts to reveal a lockable connection. The fuel line locks onto a receptacle under the flap. This prevents spills and allows the worker to attend to other jobs while filling the tank.

On the rear left side of the bus is a grating that protects the bus radiator. It is located on this side to keep it away from the dirtier, curb side of the road.

On the front of the bus you might find a bike rack. Bike riders squeeze a spring-powered handle to lower the rack into position. They lift the bike (most racks can hold only two bikes) into the steel channels that hold the wheels. A spring arm rests on top of one wheel to hold the bike in place.

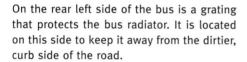

On the front of the bus are spring-loaded flaps that protect connections used when the bus is being towed or when its engine isn't running. A small round flap lifts to reveal an electrical connection so lights and other parts of the electrical system can operate without the engine running.

A larger rectangular flap protects connectors that can provide air pressure to the bus's pneumatic system. Much of a bus is powered by air: the suspension (kneeling buses), brakes, driver's seat, and doors. When the engine is not running, this is how mechanics get the doors to open.

A small flap on the right side of the front of the bus lifts to show a toggle switch. This is what the driver uses to close the door when exiting the bus.

By the front door you might find a sign indicating that the bus "kneels." To help mobility-impaired people climb onboard, the driver can lower the bus by releasing air from the suspension system on the front right.

Near the engine compartment along the back of the bus is a small rectangular flap. Beneath is an electrical connection to jump-start the bus. Rather than having to access the battery to connect the jumper cables, the mechanic can make an easier and no doubt safer connection here.

Also on the back of the bus is a speaker that blares out that "beep, beep" when the bus is backing up.

Inside the Bus

Video cameras in buses record the behavior of troublesome riders and provide a record of any actions by the driver. Interior video cameras are mounted on the inside rear wall of the bus. The recordings are recorded over if no incidents are reported.

Look at the rear and front of the bus for that small lens—you may be in the movies. Some systems also record audio.

Turn signals. Look at the steering column of a bus and one thing you won't find are turn signals. Where did they go? Check out the floor. The driver operates them by depressing either of two buttons with his or her left foot.

Accelerator and brake. Both are operated with the right foot. They are in the same relative position as in a car, but both are on the right side.

On the floor of the bus you might see square, silver-colored flaps. Beneath are attachment points for wheelchairs. The bus also carries a belt and device to latch onto the points so wheelchair riders can be secure.

Above the front door there is a glass panel. Written on the panel is a note instructing you to break the glass in an emergency to open the door. The bus doors are powered by the bus's pneumatic system. Opening the valve under the glass window releases pressure from the system and opens the doors.

Above the driver is a fire suppression system sensor. Buses have multiple sensors in the engine compartment that warn the driver if there is a fire. Dry chemical extinguisher is automatically blown into the engine compartment through several nozzles.

Bus vacuum. It opens up the doors, blows through one door, and sucks through the other to clean the bus.

Trolley

BEHAVIOR

Operates without internal combustion engines. It draws electric energy from overhead wires to power an electric motor. Some trolleys run in tracks set in the road surface. Others are buses (trackless trolleys) that use electric energy instead of diesel fuel.

HABITAT

Found in large cities. Seen frequently in Seattle, San Francisco, Boston, Dayton, and Philadelphia. They are favored especially in hilly cities.

HOW IT WORKS

A trolley pole holds conductors up against power-carrying wires supported by utility poles. Springs pull the trolley pole up and ropes, under tension by another spring, hold the end of the pole down. Pneumatic or hydraulic lifters can raise and lower the trolley pole when the bus is leaving or entering service.

The overhead lines carry high voltage (600 volts) of direct current. Carbon conductors, called shoes, act like brushes in a motor to conduct the power from the overhead wires to the wires that go to the motor.

The shoes have to be replaced every day. In winter, when ice accumulates on the wires, the carbon shoes are replaced with steel shoes that can knock the ice off the wires. But the steel shoes wear the overhead wires, so they are used just to clean the wires and then are replaced with carbon shoes. A mechanic must meet the steel-shoed bus at the end of its route and change the shoes back to carbon.

Driving a trackless trolley involves not only all the skills needed to maneuver a large vehicle in crowded city streets, it also requires the driver to anticipate turns and watch the overhead wires. If the driver drives off course, pulling the trolley pole away from the wires, the bus stops and technicians have to be called to get the bus connected again. In Seattle, the penalty for this is buying a box of donuts for the maintenance team.

Some electric buses can run on batteries or very large capacitors. Capacitors are devices that store electric energy. They now are finding more applications replacing batteries.

INTERESTING FACTS
Electric trolleys are especially popular in regions (like the Pacific Northwest) that have low electric energy costs. They are also valued in tunnels or other areas where engine pollutants are difficult to disperse.

INDEX

A

Ahearn, Thomas, 63
air bags, 38–39
air conditioning, 40–41
air filters, 122
All Wheel Drive (AWD), 54
all-terrain vehicles (ATVs),
  156
alternators, 123–124
AM/FM antennas, 14
Amphicars, 154–155
amplitude modulation
  (AM), 72–73
Anderson, Mary, 34
antennas
  AM/FM, 14
  CB (Citizen's Band), 15
  OnStar, 16
  radio, 14
  satellite radio, 17
anti-roll bars, 102
Aquadas, 154–155
automatic transmissions,
  146

B

Bagley, Rod, 90
batteries, 26, 125–126
  charging, 123
Belušić, Josip, 78
Berger, Elmer, 75
Berkeley, California, 183
bicycles
  brakes, 168
  cycle rickshaws, 172
  derailleurs, 169–170
  escalators, 165–166
  pedicabs, 172
  quick-release hubs, 171
  suspension systems, 167
Birmingham, Alabama, 34

Bishop, Arthur E., 100
boots, 115
brakes
  bicycle, 168
  cylinders, 127–128
  disc, 46, 88
  drum, 46, 88
  fluid, 46, 128
  hydraulic, 45
  lights, 44
  motorcycles, 179
  pads, 46
  parking, 68
  pedals, 45–46
  power, 45
Buchi, Alfred, 148
bumpers, 21
buses, 195
  exteriors, 199–200
  fare boxes, 197–198
  interiors, 201–202
  tracking systems, 196

C

Campagnolo, Gentullio
  (Tullio), 171
car seats, 48
Carrier, Willis, 41
catalytic converters, 89–90
Cayley, Sir George, 77
CD players, 47, 52
Charlotte, North Carolina,
  16
Citizen's Band (CB)
  antennas, 15
Claghorn, Edward J., 77
Clayton, William, 67
coil springs, 91, 104
coils, 129
combustion engines, 10
constant velocity (CV) joint
  boots, 92

convertible tops, 22–23
country codes, 21
cruise control, 49–50
cycle rickshaws, 172
cylinders, brake and
  master, 127–128

D

defrost system control, 51
Denver boots, 115
derailleurs, bicycle,
  169–170
differential gears, 93–94
dipsticks, 130
disc brakes, 46, 88
distributors, 131
drive shafts, 93
DUKWs, 157
DVD players, 52

E

Edison, Thomas, 63
Egypt, ancient, 97
electric cars, 8–9, 119–120
emergency flashers, 85
engine oil, 26
engines
  combustion, 10–11
  diesel, 118
  electric, 8–9, 119–120
  gasoline, 9
  hybrid, 120–121
  internal combustion,
    117–119
  Miller cycle, 118
  motorcycle, 182–183
  rotary (Wankel), 119
  steam, 8
  windshield wiper,
    151–152
Evans, Oliver, 8

F
fans, 132
Faraday, Michael, 41
filters
 air, 122
 oil, 135–136
flares, 53
fog lights, 25
Ford, Henry, 9–10, 156, 181
four-wheel-drive shifter,
 54–55
Fowlkes, David, 28
Freeman, Andrew, 27
frequency modulation
 (FM), 72–73
Fuchs, Sir Vivian, 159
fuel gauges, 56
fusees, 53
fuses, 57–58

G
Galvin, Paul and Joseph,
 73
gas tanks, 95
 motorcycle, 186
gasoline engines, 9
gauges
 temperature, 82
 tire pressure, 83
gearboxes, 112
gears
 differential, 93–94
 worm, 151
Getting, Ivan, 61
global positioning system
 (GPS), 60–61
glove boxes, 59
GM subscription service,
 16
golf carts, 158
Goodrich, B. F., 110
Goodyear, Charles, 110

H
halogen lights, 24
hand-cranked windows, 62

Harroun, Ray, 75
headlights, 24–25
 wiper, 34
heaters, 63
 auxiliary, 43
 block, 27
heating plugs, 26–27
High Intensity Discharge
 (HID) headlights, 25
Honold, Gottlob, 141
Hooke, Robert, 113
Hooke's Law, 91
horns, 133–134
Houdry, Eugene, 90
hubcaps and spinners, 28
hybrid motors, 120–121
hydraulic fluid, 46
hydraulic jacks, 96

I
internal combustion auto-
 mobiles, 8, 117–119

J
jacks, 96
Jackson, Wilton, 53
Johnson, Tom Loftin, 198

K
Kamen, Dean, 193
Kettering, Charles, 9, 143
key fobs, 64–65
kick sleds, 175
Kiruna, Sweden, 43

L
Lachman, Irwin, 90
Lanchester, Frederick, 88
leaf springs, 97
Lewis, Ron, 90
license plates, 29
LIDAR detectors, 71
lights
 fog, 25
 head, 24–25
Loftin Johnson, Tom, 198

M
MacMillian, Kirpatrick, 164
MacPherson, Earl, 105
manual transmissions,
 145–146
Marugg, Frank, 115
master cylinders, 127–128
Maxim, Sir Hiram Stevens,
 99
Maybach, Wilhelm, 181
Miller cycle engines, 118
mirrors
 rearview, 74–75
 wing, 35–36
Model T Ford, 9
motorcycles, 177–178
 brakes, 179
 carburetors, 180–181
 engines, 182–183
 exhaust systems, 184
 foot controls, 185
 gasoline tanks, 186
 hand controls, 187
 oil tanks, 188
 radiators, 189
 shock absorbers, 190
 sidecars, 191
motors. See engines
mufflers, 98–99, 101
Muscott, Ray H., 159

N
NASCAR race cars, spoilers
 on, 30–31

O
odometers, 66–67
oil, engine, 26
oil filters, 135–136
OnStar antennas, 16
Orukter Amphibolos, 8
Oshawa, Ontario, 16
Ottawa Electric Railway
 Company, 63

P

Parkinson, Bradford, 61
pedicabs, 172
pinion gears, 93
pistons, 11
Porsche, Ferdinand, 55
proximity systems, 19–20

Q

quick-release hubs, 171

R

rack and pinion steering, 79–80, 100
radar detectors, 70–71
radiators, 139–140
motorcycle, 189
radio antennas, 14
Radio Frequency Identification (RFID) technology, 84
radios, 72–73
rearview mirrors, 74–75
Renault, Louis, 88
resonators, 101
rickshaws, cycle, 172
roll bars, 102
Russell, James T., 47
Rzeppa, Alfred Hans, 92

S

safety wings, 31
satellite radio antennas, 17
satellites, 18
scooters, 176
seat belts, 76–77
Segways, 192–193
shifter, four-wheel-drive, 54–55

shock absorbers, 103
motorcycle, 190
sidecars, motorcycle, 191
snowcats, 159
snowmobiles, 160–161
sonar systems, 19
spark plugs, 141
speedometers, 78, 81
Spijker, Jacobus and Hendrik-Jan, 112
spinners and hubcaps, 28
spoilers, 30–31
springs, 104
starters, 142–143
steam engines, 8
steam power, 8
steering
power, 137–138
rack and pinion, 100
steering wheels, 79–80
struts, 105
sun gears, 93
superchargers, 147–148
Supplementary Restraint System (SRS), 38
sway bars, 102
Sweetland, Ernest, 136

T

tachometers, 81
tailpipes, 106
tanks, gas, 95
Teetor, Ralph, 50
temperature gauges, 82
thermostats, 144
tie rods, 107–108
tire pressure gauges, 83
tires, 109–111. *See also* wheels

radial, 110
tubeless, 110
toll transponders, 84
transfer cases, 112
transmissions, 145–146
transponders, toll, 84
trolleys, 203–204
trucks, 10
spoilers on, 31
turbochargers, 147–148
turn indicators, 85–86
on wing mirrors, 36

U

unicycles, 173–174
universal joints (U-joints), 113

V

Veeder, Curtis, 67

W

water pumps, 149
wheel clamps, 115
wheels, 114. *See also* tires
covering with hubcaps and spinners, 28
windows
hand-cranked, 62
power, 69
windshield wipers, 33–34
automatic, 42
motors, 151–152
windshields, 32
cleaning systems, 150
wipers, windshield, 33–34

Ed Sobey is an evangelist for innovative and creative learning. He gives workshops for teachers worldwide on how to teach science. Ed has directed five museums, including the National Inventors Hall of Fame, and he founded the National Toy Hall of Fame. Most recently he taught oceanography and science-teaching methods on the *MV Explorer* on a voyage around the world. He is a fellow of the Explorers Club and author of more than 20 books. Ed holds a PhD in oceanograpahy.